Flexible Automation

Flexible Automation

The Global Diffusion
of New Technology
in the Engineering Industry

Charles Edquist
and
Staffan Jacobsson

Basil Blackwell

Basil Blackwell Ltd
108 Cowley Road, Oxford, OX4 1JF, UK

Basil Blackwell Inc.
432 Park Avenue South, Suite 1503
New York, NY 10016, USA

British Library Cataloguing in Publication Data

Edquist, Charles
    Flexible automation: the global diffusion
    of new technology in the engineering industry
    1. Automation
    I. Title II. Jacobsson, Staffan
    670.42'7   T59.5

    ISBN 0-631-15971-1

Library of Congress Cataloging in Publication Data

Edquist, Charles, 1947–
    Flexible automation.

    Bibliography: p.
    1. Flexible manufacturing systems. I. Jacobsson,
    Staffan. II. Title.
    TS155.6.E27   1988   670.42   87-25606
    ISBN 0-631-15971-1

Typeset in 10/12pt CG Plantin by
System 4 Associates, Farnham Common, Buckinghamshire
Printed in Great Britain by

Contents

89 - 1311

Contents

Contents

Contents

Contents

List of Tables and Figures

Tables

List of Tables and Figures

List of Tables and Figures

Statistical Appendix Tables

List of Tables and Figures

Figures

Preface

This book deals with the diffusion of flexible automation technologies in member countries of the Organization for Economic Cooperation and Development (OECD) as well as in advanced developing countries. It is one output from a research project entitled 'Technical change and patterns of specialization in the engineering industries of India and the Republic of Korea'. The project is financed by the Swedish Agency for Research Co-operation with Developing Countries (SAREC), and therefore SAREC has also partly financed this book. In addition, the Technology Division of the United Nations Conference on Trade and Development (UNCTAD) in Geneva asked us to prepare a report on flexible automation in the engineering industry of the industrialized countries, as a part of their efforts to identify and examine policy issues of relevance to the development of the capital goods industry of developing countries. The result was published as UNCTAD TT/65 in 1985. We have also prepared another study for UNCTAD, dealing with the impact of electronics-based automation technology in the capital goods industry and its implications for developing countries. The present book draws partly on these two studies. We are grateful for the financial assistance provided by both SAREC and UNCTAD.

The bulk of the research for this book was carried out within a research programme entitled 'Technology and development' conducted at the Research Policy Institute, University of Lund, Sweden. During the work we have benefited from discussions with colleagues at the institute. Björn Elsässer has pointed to several weaknesses in the manuscript, as did three anonymous referees, while Peter Nilsson did an excellent job at the word processor. Our thanks are extended to all of them.

In late 1986 we were both appointed to positions at other universities in Sweden, Staffan Jacobsson to a research position at the Department of Industrial Management, Chalmers University of Technology, Charles Edquist to the position of professor at the Department of Technology and Social Change, University of Linköping.

Charles Edquist
Staffan Jacobsson

List of Abbreviations

AGV	automatic guided vehicles
AMBT	automatic machine building technologies
BRA	British Robot Association
CAD	computer aided design
CAE	computer aided engineering
CAM	computer aided manufacturing
CECIMO	Comité Européen de Cooperation des Industries de la Machine-Outile
CIM	computer integrated manufacturing
CNC	computerized numerically controlled
FMC	flexible manufacturing cell
FMM	flexible manufacturing module
FMS	flexible manufacturing system
JIRA	Japan Industrial Robot Association
MER	marginal export ratio
NC	numerically controlled
NCMT	numerically controlled machine tool
NIC	newly industrializing country
NMTBA	National Machine Tool Builders' Association
OECD	Organization for Economic Cooperation and Development
PC	personal computer
RIA	Robot Institute of America
SIRI	Sociata Italiana Robotica Industriale
UNCTAD	Technology Division of the United Nations Conference on Trade and Development
UNIDO	United Nations Industrial Development Organization

PART I
Introduction

1

Scope and Outline

A comprehensive process of automation has been going on in the engineering industry during the last decade or so. Various computer-based automation technologies for the design and production of engineering goods have been rapidly diffused – mainly in the industrialized countries.

Recently considerable attention has been given to the concept of computer integrated manufacturing (CIM). CIM has been called the 'factory of the future' and the 'ultimate factory technology'. CIM implies an integration of various 'islands of automation' such as robots and computer aided design (CAD). The basic premise behind CIM is to automate totally and link all the functions of the factory and the corporate headquarters. It means the integration of the whole process from the receipt of orders through design, business planning, purchasing (of components and raw materials), machining, inventory control (of parts, materials and finished products), automated warehouses, automatic vehicles, assembly, packing and marketing. All parts of the company will be influenced by such a change.

So far, however, this 'factory of the future' has remained just that. The pieces are not yet all in place. CIM is still mainly theory and conjecture, a vision of the future. One of the most difficult problems to solve is how to get disparate computers and machines to communicate with each other. For example, today the designs from a CAD system cannot normally be fed directly into a numerically controlled machine tool (NCMT). A fully integrated (CIM) factory is therefore still one or two decades away. Hence CIM has not yet had any important impact, even in the most advanced member countries of the Organization for Economic Cooperation and Development (OECD).

On the other hand, various islands of automation have certainly emerged. Much of this automation is located in the design department and on the shop floor and this kind of automation has advanced quite far in the industrialized countries and has certainly had important consequences for competitiveness, employment, skill requirements etc.

This book is devoted to a description and analysis of the diffusion of this 'island automation', i.e. to those elements of CIM which have actually started to materialize. We will focus upon NCMTs, industrial robots, flexible manufacturing systems (FMSs), and CAD systems, all essential building blocks of the factory of the future. An NCMT is a combination of mechanics and electronics and it is today considered to be a relatively mature technology which can be – and is – used extensively also by small companies. At the other end we have FMSs, which consist of machine tools, robots, automatically guided vehicles etc., all controlled by a central computer. FMSs are furthest on the path towards CIM, but they also represent an immature technology, still plagued by many technical problems. In terms of maturity, CAD systems and robots fall in between these two extremes.

The four technologies to be studied here all include electronic devices which control the operation of the machines. Together they constitute 'CAD/CAM' (computer aided manufacturing) or 'flexible automation techniques'. We will use the latter term in this study.

The recent diffusion of the four flexible automation techniques constitutes a technological breakthrough in the engineering industry which occurred after a long period of relative stability with regard to production and design technology. The most important earlier technological breakthroughs in the engineering sector were the introduction of the steam engine and the electric motor. The present shift is similar to these earlier ones in two respects. First, computers and other electronic devices will affect almost all functions and production processes and hence their impact can be expected to be quite substantial. Secondly, the process of diffusion of the electronically based technologies is not as rapid as is often believed. This means that the consequences for productivity, organization of production, quantity and quality of employment etc. have only started to materialize. Much more comprehensive effects in these respects are still to come in the next few decades.

Chapter 2 in Part I contains definitions of concepts as well as a brief presentation of a few theoretical approaches, focusing upon determinants of the diffusion of techniques. It serves as a conceptual framework for the following chapters.

Part II deals with the diffusion of the flexible manufacturing techniques in the OECD countries. First we devote one chapter to each of the four techniques (chapters 3–6). We describe the techniques, their diffusion in the OECD countries, the tasks they are used for, their industrial distribution and the types of firms using them. As far as we know, these matters are not systematically described in the existing literature in the field. This is our justification for devoting considerable attention to this descriptive aspect.

We also discuss the degree of maturity of the various techniques as well as the technical and economic reasons why the degree of maturity varies between them. This means that to some extent we also discuss the economics of the diffusion of the flexible automation techniques. In this discussion we also touch upon analysis of the determinants of the diffusion. However, we do not pursue a fully complete and systematic analysis of driving forces and obstacles to the diffusion in Part II. Neither do we try to explain differences in the degree of diffusion of the various techniques in different OECD countries.

After the descriptive chapters we address some of the implications of the diffusion of flexible automation techniques for the OECD countries. In chapter 7 we discuss the impact of diffusion of flexible automation on international competitiveness. We begin by summarizing part of the data presented earlier in an attempt to determine the impact of flexible auto-mation at an aggregated level. This impact is indicated by the share of investments in NCMTs and robots combined in total fixed investments in machinery and equipment in the engineering industry of some leading OECD countries. The impact at product and firm level is then discussed and illustrated by a detailed presentation of the use of flexible automation in two firms, one producing submersible pumps and the other, diesel engines. Finally, we compare the level of diffusion of flexible automation techniques in Japan, West Germany, Sweden, the UK and the USA and discuss the impact of the uneven level of diffusion.

In chapter 8 we discuss the employment consequences of flexible auto-mation. We show that the labour replacement effect of the four techniques is substantial at the micro level. However, we also argue that it is impossible to equate the aggregated effects of technical change at the micro level with the unemployment level at the national level.

The discussion in Part II describes aspects of the international techno-economic context within which the developing countries have to formulate their industrial and technology policies. Part III deals with the diffusion of flexible automation techniques in the Third World. This diffusion is concentrated to the more advanced developing countries, i.e. the newly industrializing countries (NICs). We therefore focus our attention mainly on these countries.

The available data on the diffusion and industrial distribution of flexible automation techniques in the NICs are first presented. Thereafter an attempt is made to analyse the determinants – i.e. the driving forces and obstacles – of this diffusion within the framework of the theoretical approaches presented in chapter 2. Hence these approaches are much more explicitly referred to here than in Part II. The analysis of the determinants includes an attempt to discuss why the NICs are lagging behind the OECD countries with regard to the degree of diffusion of flexible automation techniques. Chapter 9 is devoted to NCMTs, chapters 10 and 11 deal with the diffusion of robots, CAD and FMSs in the NICs in a similar way.

Hence, the discussion of determinants is somewhat more ambitious in Part III than in Part II. However, the discussion of the causes of diffusion is still not exhaustive and most emphasis – as in the book as a whole – is given to the description of the diffusion as such. We do not, for example, try to quantify the relative importance of various factors explaining the (differences in the) diffusion of techniques (in various countries), but pursue the discussion in a more qualitative manner, within the framework of the conceptual framework presented in chapter 2. A more ambitious quantitative analysis would require extensive and systematic micro studies, which we have not carried out.

In chapter 12 a discussion is presented of the implications for the developing countries of the technological transformation addressed in chapters 3–11. First the data on the diffusion of the various flexible automation techniques in the NICs and the OECD countries are summarized and compared. Thereafter the implications for the competitive strength in the engineering industry of the developing countries is discussed. This discussion includes the consequences not only for the NICs, but also for the poorer and less developed Third World countries. Some policy implications for the developing countries are also discussed in chapter 12.

Part IV (chapter 13) provides a review of the main findings in the book as a whole.

2
Diffusion of Technology: a Conceptual Framework

2.1 Some definitions

In this chapter we will define the concepts of technique and technology. We will also briefly present a few theoretical approaches, focusing upon determinants of the diffusion of techniques.

Technologies are often divided into product technologies and production (or process) technologies. The main concern in this book will be the latter, i.e. technologies used in the process of production.

The word 'technology' is normally used in a very comprehensive sense and often includes many important phenomena of a social character, such as knowledge, management, organization of work, other elements of social organization etc. It then becomes problematic to study the relationship between technology and social conditions, because the relation between the two phenomena cannot be satisfactorily investigated if they are not conceptually distinguished from each other (Edquist, 1985b, p. 13). For this reason we will use the term 'technique' to denote only the material elements of what is often called technology. Thus, by *production techniques* we mean tools, implements, instruments and machines which are used to produce goods and services (1985b, p. 13).

Regarding *technology*, on the other hand, we follow the tradition of using it in a more general and comprehensive sense. It includes not only technique as defined above, but also non-material elements, such as technical know-how, management, organization of work etc.

Technical change can be divided into stages even though the dividing line between the stages is not always clear and the process is not always linear. First, through research and development new ideas or *inventions* emerge, e.g. in the form of technically feasible prototypes. These are subsequently modified and developed into economically feasible techniques, i.e. *innovations*. Finally, the techniques are adopted by other producers, perhaps in foreign countries, i.e. *diffusion* is taking place. Many micro level adoptions add up to macro level diffusion.[1] It is normally not until a technique is widely diffused that it has a large economic, social and political impact. In this book we are therefore primarily concerned with the diffusion of techniques and not so much with inventions and innovations. We will describe the diffusion of flexible automation techniques in different countries. We will also discuss the driving forces behind this diffusion, obstacles retarding such diffusion and some consequences of the diffusion.

Our choice of a narrow and strictly material definition of techniques certainly does not imply that we consider the non-material elements as being of minor importance for choice, implementation and diffusion of techniques. On the contrary, these elements – e.g. know-how, organization of work, management etc. – are crucial in the process of diffusion.

2.2 Three approaches to diffusion

There is a voluminous literature on the determinants of the diffusion of techniques. Some examples are Hägerstrand (1967), Nabseth and Ray (1974), Rosenberg (1976), Rosegger (1977), Utterback (1979), Brown (1981), Gold (1981) and Rogers (1982). We will not try to summarize the findings of these authors here, but only – as a point of departure for the following chapters – briefly examine some approaches to the issues of the driving forces behind and the obstacles to the diffusion of techniques. We will present below three theoretical approaches. We want to stress that these are *not* mutually exclusive, but partly overlapping. Hence, we certainly do not pretend to present a definite theory of diffusion. Neither do we have the ambition to discuss all theoretical approaches.

2.2.1 A COMBINED STRUCTURAL AND ACTOR-ORIENTED APPROACH

The main objective of the overall activity of firms in capitalist economies is to make profits. This, of course, also governs their choices of which production techniques to adopt and use. As a consequence, the main driving force behind the diffusion of flexible automation techniques is (increased) profitability: if a computerized technique is expected to be more profitable than an 'old' one with a similar function, it will be introduced – unless there are obstacles to this. L. Nabseth has expressed this in the following way:

> In all discussions about the diffusion of new technology, the profit-
> ability or relative advantage of the new process in relation to the old
> stands out as an explanatory variable. . . . With perfect foresight and

perfect capital markets one could say that, as long as the internal rate of return on a new process exceeded a certain level (due regard being taken, of course, of the capital equipment in use), a firm ought to introduce the technique in question as soon as possible. (Nabseth and Ray, 1974, pp. 301-2.)

Rosenberg expresses a similar view in the following terms:

... the diffusion of inventions is an essentially economic phenomenon, the timing of which can be largely explained by expected profits... (Rosenberg, 1976, p. 191.)

At a less abstract level many factors may influence the profitability of adopting a new technique – or its internal rate of return. These – more specific – background factors may be of two different kinds.

1 They may have to do with *the character of the technique* as such – in relation to the old technique. The new technique may save manual or mental labour (skills); it may save capital through, for example, increased capital utilization, reduction in space requirements or increasing yield through fewer rejections. It may also save raw material. It may, of course, also save on one factor and simultaneously increase the use of another, e.g. it may be labour-saving and capital-using (i.e. require additional capital per unit of output).[2] It may lead to higher quality of the products produced. It may shorten lead time (i.e. the calendar time needed for manufacturing). The differences between the new and the old technique with respect to these characteristics may change over time when techniques are modified and improved. This is a subject to which we will return below.

2 Another set of determinants of the profitability of new techniques, and therefore of their diffusion, relates to *the structural environment* of those social actors or entities which decide to adopt new techniques – or not to do so. Factor endowments and relative prices of various production factors or the infrastructural conditions (education, credit availability etc.) may differ between countries. Institutional factors – like government policy with regard to industry and trade or legal conditions – may also vary between countries. Furthermore, the competitive structure (the degree of competition) may differ between adopting industries and it may also differ within the same industry in different countries.

The structural environment within which the actors choose techniques may also change over the course of time in the same society. The second group of factors above lead to different profitability of the same technique in different environments. Hence, the factors in this group may be particularly important in explaining differences in the degree of diffusion of a technique between countries.

There is also a third group of factors influencing the choice of techniques, but without necessarily influencing the profitability of the new technique. They are related to *the actors as such*. Large firms may be more inclined to adopt new techniques than small ones since large firms can be expected to be able to find and absorb information about new techniques quicker than small ones. Managements of different firms may also have different attitudes to (the risks involved in) choosing new techniques. Furthermore, even if the risk propensity is the same, larger firms often in fact take larger risk in an absolute sense, since the risk is smaller in relation to the total activities of the firm. Hence, on account of both a faster access to information and a greater risk-taking capacity, large firms often adopt a new technique earlier than small firms.

Some managements may also wish to adopt new techniques for status reasons or simply to learn more about them and thereby raise the technological level in the firm. Finally, the attitudes of labour unions are a consideration in these cases where they are powerful enough to influence the decisions on the adoption of new techniques.

Another aspect of the relation between the character of the actors and the diffusion process is whether the diffusion takes place within a business conglomerate or between independent firms and whether the diffusion is international or national. This may be illustrated by a four-field chart (figure 2.1). The two squares to the left concern diffusion within countries. The first possibility (1) is when techniques are diffused within a large firm with many branches in the same country and the second (2) concerns organizational diffusion but not diffusion between countries. The two squares to the right concern international diffusion, i.e. what is often called transfer of technology. In such a case the 'obstacles' to diffusion are often more severe than when diffusion within a country is concerned. Institutional

	Within countries	Between countries
Within corporations	1 Between branches of the same corporation in a certain country	3 Within transnational corporations and between countries
Between corporations	2 Between independent firms within a certain country	4 Between independent corporations in different countries

Figure 2.1 Diffusion of techniques

and cultural differences between countries may be important and the relative factor prices may be different. In addition, the information aspect may be a more severe barrier for diffusion in this case. This barrier, however, may not be expected to be as large for international diffusion within transnational corporations (3) as for international diffusion between independent corporations (4). Hence, the obstacles to diffusion may be less severe for cases in the third quadrant than in the fourth one.

The discussion above indicates that it may be important to distinguish between international diffusion within corporations and between corporations. In the former case the mechanism of international diffusion of techniques between independent corporations is ordinary trade. For the international diffusion of the non-material elements included in the concept of technology some important mechanisms are: licensing agreements, migration, scientific journals and industrial espionage.

The previous discussion implies that many of the more specific factors influencing the adoption and diffusion of new techniques may be clustered into three categories which are related to:

1 The character of the technique.
2 The nature of the structural environment of the actor.
3 The characteristics of the actors themselves.

To sum up, the diffusion of techniques may be highly influenced by structural factors, i.e. the choice of technique is not only, or even primarily, a matter of conscious policy-making by decision-makers who function as actors. At the same time social actors do make the decisions to implement some techniques rather than others, and they often have some freedom of action. Therefore a combination of a structural and an actor-oriented approach is useful in a study of diffusion of techniques.[3]

2.2.2 THREE PHASES OF DIFFUSION

The discussion in this chapter has hitherto been mainly of a static nature. Changes in the rates of diffusion of techniques *over time* are often illustrated with the help of so-called S-curves. Time is depicted on the horizontal axis and some measure of diffusion (e.g. proportion of firms using the technique) on the vertical axis.[4] The curve may be divided into three phases. In the first – introductory – phase (A), diffusion is slow. The second phase (B), where the curve is steep, is often called the diffusion or growth phase. Finally, we have the saturation or maturity phase (C), which may later be transformed into a decline phase. The shape of the S-curves varies, of course, between techniques and countries (figure 2.2).

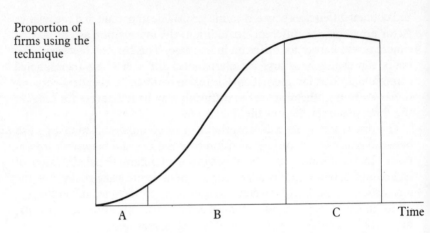

Figure 2.2 The S-curve

The determinants behind the movement of the technique along its S-curve
are complex and include many factors. Two important determinants will
be further discussed here. They are changes in the *supply side* and in the
diffusion of *information*.

Rosenberg (1976, pp. 190–210), in his analysis of factors affecting the
diffusion of technology, argues that 'factor and commodity prices aside,
the rate at which new technologies replace old ones will depend upon the'
speed with which it is possible to overcome an array of supply side problems'
(p. 191). We will now discuss further these supply side problems, which
often serve as obstacles to the diffusion of new techniques.

In the *introductory* phase of an S-curve, the product or the technique is
often produced in small quantities and for very specific customers. Utterback
(1979, p. 47) suggests that 'the initial uses of a major product innovation
tend to be in small...market niches in which the superiority of the new
product...allows it to command a temporary monopoly, high prices and
high profit margins per unit.' Often, the new product or technique is
characterized by functional superiority rather than by lower price. In this
phase, close contact with the market is also necessary for the producer in
order to acquire ideas for product developments. The new machine is often
technically immature and very costly in relation to the machine and/or
worker it is to replace (Jacobsson, 1986, chapter 3). Indeed Rosenberg (1976)
argues that:

> ...if one examines the history of the diffusion of many inventions,
> one cannot help being struck by two characteristics of the diffusion
> process: its apparent overall slowness on the one hand and the wide
> variations in the rates of acceptance of different inventions, on the

other.... a better understanding of the timing of diffusion is possible by probing more deeply at the technological levels itself, where it may be possible to identify factors accounting for both the general slowness as well as wide variations in the rate of diffusion. (p. 191.)[5]

As the product moves into its *growth* phase, changes take place both in the technique itself and in the type of adopters. Outside the large firms that are normally the first adopters, there are, particularly in the engineering sector, a very large number of medium-sized and smaller firms. These firms often demand a somewhat different technique than the larger firms. Generally speaking, these medium and smaller firms have less in-house technical expertise and are therefore forced to rely more on mature technologies. They are smaller and, in general, therefore cannot profitably use as much specialized machinery as the larger firms, they need more universal and flexible machines. They are also more price conscious. Finally, their smallness also means that economies of scale in *buying* the new technique (in terms of information search) and in *using* it (in terms of specialized skills) have to be reduced before they can buy the new technique.

The different demands made by the two broad groups of firms in the engineering sector mean that the product or technique needs to be altered if it is to move along the S-curve into the growth phase. The growth phase can thus be characterized by the penetration of new markets through:

1   Standardization of the technique, i.e. less custom-designed features and less need for specialized skills to use it.
2   Product differentiation, away from the specifications set by the larger customers, which broaden the range of models available to include also more simple types.
3   Price reductions as a consequence both of differentiation into more simple models and of the reaping the benefits of economies of scale in the supplying industry (Jacobsson, 1986, chapter 3).

Thus, the movement of a technique along the S-curve can be seen at least partly as a function of the supplying industry's behaviour in that it changes the technical characteristics of the product to make it more attractive to a larger market, and produces it in large enough quantities to be able to reduce its relative price. The supplying industry thus changes the characteristics of the technique and its price. Such progressive modifications of techniques are also stressed by Gold (1981, p. 248). They may mean that the maximum potential level of diffusion gradually increases over time.

Another determinant of the speed at which a technique moves along the S-curve is the spread of information about the technique. Nabseth and Ray (1974, pp. 6, 229–301), as well as Rogers (1982), emphasize this aspect.

Another example is Hägerstrand (1967), although he does not deal with the diffusion of industrial process techniques, but with product technologies and agricultural process technologies (p. 12). His conceptualization of the diffusion process has been summarized by Brown (1981) in the following way:

> The basic tenet of Hägerstrand's conceptualization of the spread of innovations across the landscape is that the adoption of an innovation is primarily the outcome of a learning or communication process. This implies that factors related to the *effective* flow of information are most critical and, therefore, that a fundamental step in examining the process of diffusion is identification of the spatial characteristics of information flow and resistance to adoption. (p. 19.)

Hägerstrand treats innovation diffusion as an information diffusion problem. We certainly agree that information about a new technique is a necessary condition for the diffusion of it; but it is certainly not a sufficient one.[6] Of course, the diffusion of information is not independent from the behaviour of the supplying industry. Through international marketing and the setting up of repair and maintenance services at home and abroad, a considerable amount of information about the technique may be diffused.

The fact that information about a technique is a necessary condition for its diffusion is sometimes neglected by neoclassical economists. One example is Rosegger (1977). A reason for this neglect may be that 'the prevailing assumption of micro theory is that decision-makers fully "know" the characteristics of all techniques available at any given moment in time' (Rosegger, 1977, p. 91).

In the *maturity* phase of the S-curve, the technique has reached its potential level of diffusion and the situation is stable until a new technique starts to replace it.

2.2.3 SOCIAL CARRIERS OF TECHNIQUES – AN INTERDISCIPLINARY APPROACH

Let us conclude this chapter by presenting an interdisciplinary approach to determinants of the diffusion of techniques. It takes into account several of the relevant factors discussed earlier in this chapter.

Diffusion of new process techniques means that they are chosen by (additional) actors as a substitute for 'old' techniques fulfilling the same or similar productive functions. The process of diffusion of techniques is not only an economic process; it is also determined by social and political factors. In earlier work (Edquist and Edqvist, 1939; Edquist, 1985b) the concept of 'social carriers of techniques' was formulated with the intention to capture a broad range of determinants of the choice of techniques. We will begin by defining the general concept of 'social carriers of techniques'.

A social carrier of a technique is a social entity which chooses and implements a technique; it 'carries' it into the society. It is defined in the following way. For a certain technique to be chosen and implemented in a specific context or situation, the technique must, of course, actually *exist* somewhere in the world, i.e. it must be 'on the shelf'. But some additional conditions must also be fulfilled:

1   A social entity that has a subjective *interest* in choosing and implementing the technique must exist.
2   This entity must be *organized* to be able to make a decision and also be able to organize the use of the technique properly.
3   It must have the necessary social, economic and political *power* to materialize its interest, i.e. to be able to implement the technique chosen.
4   The social entity must have *information* about the existence of the technique and functionally similar ones.
5   It must have *access* to the technique in question.[7]
6   Finally, it must have, or be able to acquire, the necessary *knowledge* about how to handle, i.e. operate, maintain and repair, the technique (Edquist and Edqvist, 1979, pp. 31–2, Edquist, 1985b, 77).[8]

If the six conditions listed above are fulfilled, the social entity is a *social carrier of a technique*. The carrier may be, for example, a private company, an agricultural cooperative or a government agency.[9] Every technique must have a social carrier in order to be chosen and implemented. If the six conditions are simultaneously fulfilled, the technique will actually be introduced and used. In other words, the six conditions are not only necessary but, taken together, they are also sufficient for implementation to take place (Edquist and Edqvist, 1979, p. 32; Edquist, 1985b, p. 77).[10]

This has certain implications. If, for example, one of the six conditions is unfulfilled, it can be said to constitute an obstacle to the adoption of the technique in question. It then constitutes a constraint for the diffusion of the technique. In policy of supporting the diffusion of new techniques, it thereby constitutes a point of intervention. Hence, an analysis of this kind may have certain policy implications. For example, an actor with an interest in adopting a certain technique and with ambitions to become its social carrier must of course, in its strategy or policy to achieve this, concentrate upon those conditions among the six defining a social carrier which are not fulfilled. These may, of course, be different for various actors, but also for similar or corresponding actors situated in different structural environments. In a strategy, the missing conditions must be identified by the policy-maker and the means to overcome these must be sought (Edquist, 1985b, pp. 100, 105–6).[11]

The approach outlined above – just like the one presented in section 2.2.1 – is a combination of a structural and an actor-oriented perspective.[12] Many of the determinants of diffusion discussed previously also reappear in the six conditions defining a social carrier of a technique, but in a more general – and perhaps also more systematic – manner. The approach incorporates economic as well as social and political factors. As a result of the very nature of the process of diffusion of techniques, such an inter-disciplinary approach may be advantageous. However, the economic factors are often the more important ones, at least in capitalist countries.

This study is confined to the diffusion of flexible automation techniques in capitalist market economies. This means that the decisions to choose these new techniques instead of non-electronic ones are mainly taken by manage-ments of private companies (in some cases in public companies also); they are the main actors in the process of diffusion of the techniques.[13] Let us now very briefly try to relate the six conditions to the choice of flexible automation techniques of firms in capitalist market economies.

1 The *interest* condition is, of course, in this case the profit motive, which is generally agreed upon as being a major driving force behind diffusion of techniques in capitalist countries. In section 2.2.1 we discussed some background factors – the character of the technique and the structural environment – which influence the profitability for firms of adopting new techniques.

2 The condition of *organization* is normally fulfilled by capitalist firms, although the introduction of flexible automation techniques may certainly sometimes require comprehensive organizational changes.

3 The *power* condition is also normally fulfilled by capitalist firms. However, legal rules may prohibit the use of certain techniques. Trade union activity may also prevent firms from adopting them, i.e. there may be conflicting interests within firms.

4 The condition of *information* is – as we have seen above – often strongly stressed as being very important in the literature on the diffusion of techniques.

5 The condition of *access* may be unfulfilled simply because of lack of financial resources to buy the technique. Secrecy, patents as well as embargoes and other political constraints may also restrict access to advanced techniques in some cases.

6 The condition of *knowledge* may be a substantial obstacle to the adoption of new techniques in all countries. As we will see later on, this problem may be particularly important for the introduction of flexible automation techniques in developing countries. It is an example of the crucial importance of the non-material elements in the definition of the concept of technology.

In the following chapters we will consider the determinants of the diffusion of flexible automation techniques. We shall, however, concentrate on the description of diffusion as such and be less ambitious in respect of causal analysis. We will not, for example, try to quantify the relative importance of various factors explaining (differences in) the diffusion of the techniques in various countries. We will rather pursue the discussion in a more qualitative manner within the conceptual framework presented above and using the theoretical elements presented in this chapter as a point of departure.[14] With regard to the OECD countries, this discussion will be more implicit than in the case of developing countries.

Notes

1   A similar division into stages can be made with regard to technological change if, for example, organizational changes are included in the discussion.

2   To determine the factor-saving bias of a new technique in relation to an old one presupposes an assumption that relative factor prices are given. The relative factor prices form part of the structural environment of those actors choosing techniques.

3   However, it is not possible to determine in general or a priori the relative weight that should be given to each of the perspectives. Such a judgement can only be made on an empirical basis and for specific cases, since the structural constraints vary, to a considerable extent, in different particular instances (Edquist, 1980, p. 23; Edquist, 1985b, pp. 10–13).

4   Diffusion of a technique may be measured in various ways. Two possibilities are (i) proportion of the firms (in an industry) which use it; (ii) the share of output, capacity or employment of labour which it accounts for in relation to the industry's total output, capacity or employment (Nabseth and Ray, 1974, p. 8). The appropriateness of the various measures varies with the objective of the measuring. The most difficult task is often to determine the potential diffusion of a technique, i.e. the denominator in the various fractions.

5   In the rest of the article Rosenberg addresses a number of factors that may serve as obstacles to the diffusion of new techniques, particularly in the introductory phase of the S-curve. They are:

• Most inventions are relatively crude and inefficient at the date when they are first recognized as constituting a new invention and improvements are therefore necessary before rapid diffusion.
• The development of the human skills on which the use of the new technique depends in order to be effectively exploited.
• The skills and facilities in machine-making itself.
• The complementarity in productive activity between different techniques.
• The 'old' technique continues to be improved.
• The institutional context can account for the very slow diffusion of a superior technology (Rosenberg, 1976, pp. 195–210).

6 All processes of social, economic and technical change are multicausally determined. This includes, of course, diffusion of techniques. Therefore it is always incorrect to talk about *the* cause of technical change or diffusion. Hence, a study of determinants of a certain choice of technique is a question of identifying all – or the most important – factors that result in a certain choice of technique, i.e. causally to explain a multicausally determined process (Edquist, 1985b, p. 12).

7 If the condition of information is fulfilled for an existing technique, this does not mean that the technique is available to the social entity. The latter must also be able to gain access to the technique in a physical sense, e.g. by purchasing it. Therefore, it is useful to treat information and access as separate conditions, although they are partly overlapping (Edquist, 1985b, p. 104).

8 The concept of social carriers of techniques was theoretically developed in a rationalistic manner and defined in Edquist and Edqvist (1979). It was marginally modified in Edquist (1985b). In Edquist (1985b) it was also used in a detailed theory-based empirical study of determinants of the choice of technique in sugar cane harvesting in Cuba and Jamaica from the late 1950s to the early 1980s (Edquist, 1985b, pp. 79–106).

9 'Social carriers of techniques' are specific kinds of actors. The intention is that this 'technique-centred' social concept – and the six conditions defining it – shall function as a conceptual bridge, or intermediary link, between the structure of society and technical change, to facilitate a detailed analysis of the interaction between techniques on the one hand and socioeconomic and political conditions (in a wide sense) on the other (Edquist, 1985b, p. 77).

10 This is analytically true, and not an empirical hypothesis. The empirical work concerns determining when the various conditions are fulfilled for which actors and in which structural contexts (Edquist, 1985b, p. 104). Such an analysis was carried out in Edquist, 1985b, ch. 6.

11 The first three conditions defining a social carrier of a technique refer to characteristics that are inherent to the social entity. Interest, organization and power imply the actual constitution of the social category as such. The latter three conditions have a different character. They are related specifically to the technique. If the first three conditions are fulfilled, information, access and knowledge can often be acquired from the outside world by the social entity – through literature, trade, experts, advice through consultants etc. (Edquist and Edqvist, 1979, p. 33).

12 The 'social carriers' approach may seem to be exclusively actor-oriented. However, the distribution of power among the actors in the society is affected by structural factors. The interests of various actors are also affected by structural conditions like factor endowment and factor prices. Hence, some of the conditions defining a social carrier are directly associated with structural phenomena. The concept of social carriers of techniques is therefore intrinsically 'structure-based' in the sense that it is defined, in part, from a structural point of view. This means that the 'social carriers' approach is less actor-oriented than it may seem at first glance (Edquist, 1985b, pp. 78, 104).

13 Other relevant actors are government agencies and labour unions. A complex interplay between firm managements, unions and governments may influence the

diffusion of new techniques. This interplay may be harmonious or characterized by conflicts – depending on the interest of the actors and the distribution of power between them.

14 An ideal quantitative causal explanation would require extensive and systematic micro studies, which we have not carried out.

PART II

The Diffusion of Flexible Automation Techniques in the Engineering Industries of the OECD Countries

Part II

3

Numerically Controlled
Machine Tools

3.1 The Technique

Within the engineering industry the machining function is central to the
production process. Table 3.1 shows that about 20 per cent of the time of
blue collar workers in the Swedish engineering industry in 1981 was spent
on machine tool operation. Another 10 per cent is expended on toolmaking,
setting and repair and maintenance, tasks intimately linked to the process
of cutting and forming metal. It has been estimated that there are some
3,000 different types and sizes of machine tools (Machine Tool Trades
Association, 1983). A very broad classification would be to distinguish
between metal-cutting and metal-forming machine tools. The former
accounted for 78 per cent of the stock of machine tools, in units, in the
USA in 1983 (*American Machinist*, 1983a).

For a very limited number of products, e.g. engine blocks, the production
volumes have – during recent decades – justified the investment in rigid,
special purpose automatic production systems, e.g. transfer lines. The bulk
of engineering products, however, are produced in small and medium
batches. Indeed one source suggests that in Japan this type of production
accounts for 70–80 per cent of the value of production (Nakao, 1983,
p. 743). The workshops catering for a diversified demand, e.g. 1,500 types
of pump, must have a very flexible production apparatus. The need for
flexibility meant, until recently, that multi-purpose and hand-operated
machine tools were used. It was thus not possible for entrepreneurs to benefit
from automation in the bulk of the engineering industry.

A number of different tasks can be identified in the operation of a machine
tool:

1 The workpiece is transported to the machine.
2 The workpiece is fed into the machine and fastened.

Table 3.1 Hours worked by major categories of blue collar workers in
the Swedish engineering industry in the first quarter of 1981
as a percentage of the total

Worker category	%
Assembly workers	21.3
Machine tool operators	20.2
Quality control personnel	7.8
Warehouse workers	7.6
Welders	7.0
Transport and cleaning workers	5.1
Instructors and setters	4.4
Tool makers	3.8
Platers	3.1
Painters	2.7
Polishers and other surface treaters	2.3
Repair workers	2.2
Foundry workers	1.2
Other workers	11.1
Total	99.8

Source: Elaboration on the Joint Wage Statistics of the Union of Swedish Metalworkers and
the Association of Swedish Engineering Industries.

3 The right tool is selected and inserted into the machine.
4 The machine is set, e.g. operation speed is determined.
5 The movement of the tool is controlled.
6 The tool is changed.
7 The workpiece is taken out of the machine.
8 The workpiece is transported to another machine tool or to a warehouse
   or to assembly.
9 The whole process is overlooked in case of tool breakages, etc.

The first numerically controlled machine tool (NCMT) was developed in
the 1950s. Instead of having a worker perform tasks (4) and (5) above, the
information needed to produce a particular part was put on a medium, e.g.
a tape, and fed into a numerical control unit. By simply changing the tape,
the NCMT could quickly be switched from the production of one part to the
production of another. Flexibility and automation were combined. Because
of the high costs of the NCMTs and the unreliability of the numerical
control unit, the technology was not diffused widely until the early 1970s
when the numerical control unit began to be based on minicomputers. A still

more significant change in the technology was the introduction of micro-computers as the basis for the numerical control unit, a process which began about 1975. The use of microelectronics was associated with an increase in reliability, a simplification in programming and the automation of other tasks, in addition to (4) and (5). Tool changing is normally automatic today – tasks (3) and (6) – and automatic material handling equipment is supplied by the leading firms in the industry, automating tasks (2) and (7). Finally, the essential task of overlooking the production process – task (9) – has begun to be automated through automatic diagnosis of faults etc.

3.2 The Diffusion of Numerically Controlled Machine Tools

The rapid diffusion of NCMTs began at the end of the 1970s. Today it is extensive. In 1983 the stock of NCMTs in the USA alone amounted to 103,000 units (*American Machinist*, 1983a). In table 3.2 the share of NCMTs in total machine tool investment is shown to have increased between 1978 and 1984 in some OECD countries. In the case of Japan, there was a rise from 15.6 per cent in 1978 to 54.3 per cent in 1984. In the case of the UK, the share of NCMTs rose from 19 per cent in 1978 to nearly 41 per cent in 1982, only to rise to 62.4 per cent in 1984!

Table 3.2 Share of NCMTs in total machine tool investment in Sweden, the UK, Japan and the USA, 1978–84, as a percentage in monetary terms

Year	Sweden[a]	UK	Japan[a]	USA
1978	26.0	19.0	15.6	n.a.
1979	31.1	22.5	27.2	n.a.
1980	28.6	30.9	28.3	27.8
1981	30.6	44.9	29.3	30.2
1982	31.4	40.8	38.8	38.1
1983	55.0	54.6	47.5	43.8
1984	59.4	62.4	54.3	40.1

[a] Investment in metal-forming NCMTs is not included due to non-availability of data for Japan (all years) and for Sweden 1978–82.

Sources: Sweden – for 1979–82, elaboration on data supplied by the Swedish Machine Tool Builders' Association. For 1978, Computers and Electronics Commission, 1981, p. 173. For a discussion of the methods used to estimate the 1983 and 1984 data, see n. 1. UK – elaboration on Machine Tool Trades Association and *Metalworking Production*, 1980, 1981, 1983, 1985). USA – elaboration on National Machine Tool Builders' Association (NMTBA), 1981/2, 1982/3, 1984/5. Japan – elaboration on data supplied by the Japan Machine Tool Builders' Association, CECIMO, NMBTA, 1984/5, and Japan Tariff Association, 1985.

The introduction of numerical control varies tremendously, however, depending on the type of machine tool. In 1976, only 25 per cent of the value of production of metal-cutting machine tools in the largest OECD countries consisted of NCMTs, a share that rose to 41 per cent in 1982 and to 56 per cent in 1984. For metal-forming machine tools,[2] the share of NCMTs was only 3 per cent in 1976, 12 per cent in 1982 and 19 per cent in 1984. Thus, NCMTs are primarily used for metal cutting as opposed to metal forming (see appendix table 3.1).

Among the metal-cutting machine tools, it is those performing turning operations (lathes), milling, drilling and boring operations which have been most affected by the developments in electronics. As shown in table 3.3, the share of NCMTs in the production of machine tools performing these functions was already 36 per cent by 1976 and had grown to 66 per cent in 1982 and to 76 per cent in 1984. One may note here that there was an absolute decline in the (current) value of production of conventional boring, milling, drilling, gear-cutting machinery, lathes, punching and shearing machines and presses (see appendix tables 3.1 and 3.2).

Table 3.3   Share of NCMTs in total production of milling, drilling and boring machines, lathes and machining centres in six OECD countries,[a] 1976, 1982 and 1984

	1976		1982		1984	
	(US$m)	*(%)*	*(US$m)*	*(%)*	*(US$m)*	*(%)*
NCMT	1,145	36	3,658	66	3,750	76
Conventional	2,005	64	1,846	34	1,157	24
Total	3,150	100	5,504	100	4,907	100

[a] USA, Japan, Federal Republic of Germany, France, Italy and UK.
Source: Elaboration on data supplied by CECIMO.

NCMTs for turning, boring, drilling and milling functions can be said to be mature technologies in terms of the S-curve concept. They had already moved to the stage in their product cycle where standardization and mass production was essential by the latter half of the 1970s. Today, the main innovative efforts lie in system building, which we will deal with in chapter 5. The mature character of the technologies is reflected in the diffusion of these NCMTs to smaller firms too. In Japan, firms with less than 300 employees have accounted for the majority of sales for some time (see appendix table 3.3). In terms of stock, data from the USA in 1983 show that 40 per cent of the number of NCMTs installed are in firms with less than 100 employees (see appendix table 3.4). In the period 1978–83, these firms accounted for 47 per cent of the market for NCMTs in the USA.[3]

If we add to this market share approximately 15 per cent[4] from the group of firms with 100–300 employees, the share of 'medium and small firms' in the market for NCMTs would be about the same as that in Japan in 1980–1. Hence, also in the USA the smaller and medium-sized firms have accounted for the bulk of the market for some time.

The process of maturation of NCMTs, in particular computerized numerically controlled (CNC) lathes and machining centres (a combined milling, drilling and boring machine), was intimately connected to the behaviour of the supplying industry.[5] In the early 1970s, the supplying industry had, as a rule, not yet identified NCMTs as the key product around which they should define their strategies. Although there was some production and sales of NCMTs, the business relations between the NCMT producer and the buyer were mainly of a local or regional nature. The volume of production of each NCMT producer was low and the main customers served were the larger firms. These large firms, moreover, normally demanded high performance machines, often with custom-designed features.

In the mid 1970s some Japanese firms started to apply a strategy which could be labelled an 'overall cost leadership' strategy.[6] The firms had as a basic objective to penetrate a very large part of the engineering industry and this was achieved by differentiating their CNC lathes from the traditional suppliers in the other OECD countries; by a standardization of these CNC lathes; by an increase in the production volumes which thereby enabled them to benefit from various sources of economies of scale. In terms of the S-curve terminology, these firms created the point of maximum curvature and made the industry move upwards on the S-curve.

The key factor involved in the definition of strategy was the design of lower performance, smaller and lower cost CNC lathes than had hitherto been the case. These lower performance and standardized CNC lathes were primarily aimed at the smaller and medium-sized firms. In other words, these Japanese firms deliberately tried, and succeeded in, opening up the large market of smaller and medium-sized firms which, as was argued in chapter 2, often demand a slightly different type of technique than larger firms. On the other hand, the European and US producers, very broadly speaking, took the production problems in larger firms as the point of departure for their design efforts.

The difference between the type of NCMTs produced in Japan and in the other OECD countries can be illustrated by the large discrepancies in average weight per CNC lathe produced. Whilst in 1984 the average weight was 4.2 tons in Japan, it was 8.4 tons in the Federal Republic of Germany (see appendix table 3.5).

The success of the Japanese firms in opening up the new market of smaller and medium-sized firms allowed them to capture significant economies of scale. This meant that the size of the leading firms grew in a phenomenal way in the period 1975–82 (see appendix tables 3.6 and 3.7). At least some of these benefits were passed on to the consumers in terms of lower prices on CNC lathes.[7]

Hence, the behaviour of the supplying industry in the case of NCMTs, in the forms of the implementation of the 'overall cost leadership strategy' by some Japanese firms, had an immediate influence on the actual definition of the S-curve and on the process of maturation of the technique.

3.3 The Industrial Distribution of Numerically Controlled Machine Tools

Table 3.4 shows the distribution of NCMTs by machinery groups in Japan (1981) and in the USA (1983). The general machinery sector, broadly ISIC 382, accounts for approximately half of the installations. The transport equipment industry is the second largest user of NCMTs.

Table 3.4   Distribution of the stock of NCMTs by sector in Japan (1981) and the USA (1983)

	Japan[a]		USA[a]	
	(no.)	(%)	(no.)	(%)
General machinery	11,394	43	52,541	51
Electrical machinery	4,262	16	10,772	10
Transport equipment	6,276	23	15,284	15
Precision machinery	1,775	7	4,874	5
Metal products	1,460	5	14,463[b]	14
Casting/forging products	580	2	2,662[c]	3
Miscellaneous	978	4	2,712	2
Total	26,725	100	103,308	100

[a] The Japanese inventory covers plants with 100 employees and more. The USA inventory covers all size classes.
[b] Fabricated metal products
[c] Primary metals.
Sources: Elaboration on *Metalworking, Engineering and Marketing*, September 1982, p. 75; *American Machinist*, 1983a, p. 118–19.

Data at a more disaggregate level exist for the USA in 1983. Appendix table 3.8 shows the number of NCMTs installed at branch level as well as the value added in 35 branches in the US engineering industry. From the table, we can observe that the concentration of NCMTs to a few branches is very high. The five largest users of NCMTs account for nearly 50 per cent of the units installed. These branches are:

1 Miscellaneous machinery, except electrical, which means mainly what are called 'jobshops'.
2 Metalworking machinery.
3 Aircrafts and parts.
4 Construction, mining and material handling machinery.
5 General industrial machinery, e.g. pumps, compressors etc.

However, the greatest impact – on productivity etc – is not necessarily in the same branches, since the size of the value added as well as the importance of machining in the production process varies significantly between the branches. In figure 3.1 we have therefore mapped out the intensity in use of NCMTs in the different branches in the USA in 1983. The intensity is measured by two indicators, each being an axis in the diagram. The vertical axis shows the normalized share of NCMTs in the total stock of machine tools in the various branches (column 7 in appendix table 3.8). This indicator shows the likelihood of choosing NCMTs instead of other machine tools for each branch. On the horizontal axis we have indicated the ratio between the number of NCMTs installed and the value added (measured in billion US$) in the branches (column 6 in appendix table 3.8). This ratio is a rough indicator of the importance of NCMTs in the value added process in the different branches. Thus, the further 'north-east' we move in the diagram, the greater is the impact of NCMTs.

Eight branches fall in the north-east quadrant of figure 3.1. These eight branches are therefore those which would feel the greatest impact of NCMTs, according to both indicators of the intensity of use of these machine tools. We may observe that among these eight branches are the five branches, mentioned above, which are the largest users of NCMTs. Thus, apart from these branches, we need to add three to the list of branches that are most affected by this technique. These are special industrial machinery, engines and turbines and miscellaneous transportation equipment.

The branches found in the north-west quadrant – a total of four – have a relatively high intensity in use of NCMTs as measured by the share of these machine tools in the total stock of machine tools. On the other hand, the intensity in use of NCMTs, as measured by the number of NCMTs divided by value added in the respective branches, is low. This indicates that machining is frequently done with NCMTs in these branches but that machining (in general and with NCMTs) plays a relatively minor role in the production process. Not surprisingly, three of the four branches are based on electronic technologies as distinct from mechanical technologies. In the south-east quadrant we find only three branches, screw machine products, ordnance accessories and miscellaneous fabricated metal products. Screw machine products in particular are produced in very large series, which

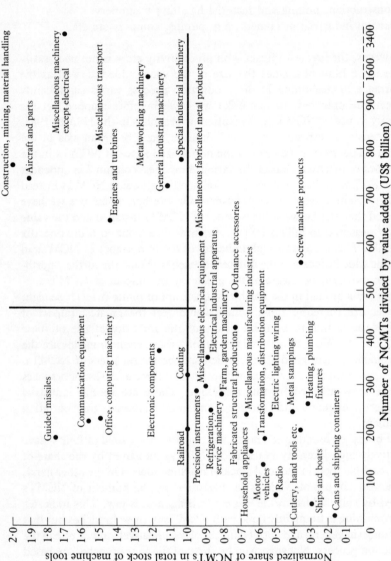

Figure 3.1 The intensity in use of NCMTs by branch in the USA in 1983 as measured by normalized share of NCMTs in total stock of machine tools and number of NCMTs installed divided by value added.

suggests that non-flexible automation is used more than NCMTs. Machining with NCMTs is, however, still a relatively important part of the value added process, probably due to the great importance of machining in the production process. Finally, in the south-west quadrant we have all those branches which are not so greatly affected by NCMTs.

The distribution of NCMTs between the various branches can, of course, be explained to a considerable extent by the nature of the production process. In table 3.3 we saw that numerical control was to a very large extent applied to machine tools performing the turning, boring, drilling and milling functions. Indeed, in 1983, 84 per cent of the stock of NCMTs in the USA were machines carrying out one or several of these functions (*American Machinist*, 1983a). We would, of course, expect to find a wide diffusion of NCMTs in industries where these functions play an important role in the value added process. In order to assess the relationship between the diffusion of NCMTs and the importance of these functions an analysis has been made for Sweden.

We correlated (a) the branchwise share of the total stock of NCMTs and (b) the branchwise share of the number of hours worked by workers operating milling, drilling, boring machines and lathes. The correlation coefficient proved to be 0.82 and the coefficient of determination therefore 0.67. Thus, the pattern of diffusion of NCMTs can to a very large extent be explained by the nature of the industrial process at branch level. Still, since these coefficients are not closer to 1, other factors explaining the choice between conventional machine tools and NCMTs at the branch level must also exist. In the aerospace industry, where extremely high quality is of over-riding concern, there is a strong preference for NCMTs. In construction, mining and material handling machinery, machine tools and general industrial machinery, there is a great demand for flexibility which again leads to a preference for NCMTs. In contrast, some automobile components have, traditionally, been produced in such large quantities that the flexibility of NCMTs has not been crucial.

3.4 The Economics of Numerically Controlled Machine Tools

The economic efficiency of (stand-alone) NCMTs is well established. The degree of saving on the cost of production of a part varies, however, as would be expected, from case to case. In one study of the Federal Republic of Germany (Rempp et al., 1981), the range of total cost saving varied from 3 per cent to 40 per cent. In a Swedish study of six firms (Elsässer and Lindvall, 1984, ch. 8), five firms decreased their cost of production of parts by using NCMTs (though for the sixth firm the use of NCMTs resulted in an increased total cost of production). The maximum cost reduction amounted to around 50 per cent.

Let us illustrate the economies of the choice between conventional machine tools and NCMTs with an example. Table 3.5 shows an investment calculation where the choice was between using CNC lathes or conventional lathes. The calculation is based upon cost data collected from a Swedish firm producing standardized pumps.

Table 3.5 Investment calculation comparing CNC and conventional lathes, assuming one shift operation (Sw.kr. per unit of output)[a]

Cost item	CNC lathes	Conventional lathes
Machine investment p.a.	875,000	560,000
Building		126,000[b]
Labour	793,968	2,101,680
Total costs	1,668,968	2,787,680

[a] Other assumptions are: (1) One CNC lathe operator's annual wage (including social security and other costs incurred by the employer) is 96,744 Sw.kr. (2) A setter's wage is 116,760 Sw.kr. One setter is used for seven CNC lathes. (3) One conventional lathe operator's wage is 100,080 Sw.kr. (4) The p-factor is 3, i.e. three conventional lathes are assumed to produce as much as one CNC lathe. (5) Ten year depreciation and 15 per cent interest rate.
[b] Only the *extra* cost of buildings (i.e., for more space) associated with the choice of conventional lathes is known.
Source: Jacobsson, 1986, ch. 2.

The manager of the firm took costs of investment and labour into account in his calculation. He also included savings in space (i.e., cost of building) from using CNC lathes instead of conventional lathes. He calculated with a saving of 30 square metres saved per machine. The reason for saved space is simply that three conventional machines take up more space than one CNC lathe.

Given the assumptions of a depreciation time of ten years, an interest rate of 15 per cent and one shift operation, total production costs per unit of output in the case of CNC lathes were only 60 per cent of the case in which conventional machine tools were used. Using the CNC lathes implied an increase in investment costs, by 27 per cent, and a large decrease in labour cost, by 62 per cent.

If, however, it is assumed that the CNC lathes are used in two shifts and the conventional lathes in one shift, the results are, of course, different (see table 3.6). This situation is common in a country like Sweden where operators of conventional machine tools normally work only one shift whilst

NCMTs are operated in two or even three shifts. This difference is a function of the skill content of the operators' work and the relative strength of the unions *vis-à-vis* the employers. Total costs per unit of output with CNC lathes in this example were only 52 per cent of those of conventional lathes. It should also be noted that the use of CNC lathes results in a reduction in investment costs per unit of output by nearly 16 per cent. Labour costs are reduced by 58 per cent.

Table 3.6 Investment calculation assuming two shifts for CNC lathes and one shift for conventional lathes (Sw.kr. per unit of output)[a]

Cost item	CNC lathes	Conventional lathes
Investment p.a.	1,041,000	1,235,000[b]
Labour	1,587,936	3,783,024
Total costs	2,628,936	5,018,124

[a] A depreciation time of 7.5 years is assumed for CNC lathes. It is also assumed that the output of the second shift for CNC lathes is only 80 per cent of that of the first shift. The other assumptions are the same as in table 3.5.

[b] The extra costs for the building are included in investment costs p.a.

Source: Jacobsson, 1986, ch. 2.

Thus, the main source of savings from the adoption of NCMTs is the increased labour productivity. Labour cost per unit of output decreased in this example by around 60 per cent. Pump production is, however, one of the most advantageous applications of CNC lathes and often labour cost per unit of output declines less than in this case.

As the example showed, NCMT adoption can either increase or save on investment costs per unit of output, depending upon the shiftwork practice. Another factor determining changes in the capital/output ratio is the cost of work in progress and stocks. To the extent that the use of NCMTs can reduce the need for stocks (by increasing the flexibility of the production apparatus in relation to more fixed automation or by increasing the throughput of parts), interest payments caused by stocks can be reduced. The importance of this factor varies naturally between firms as the relative values of parts, components and final products differ. In the pump example, the manager did not include this aspect in his analysis since the pumps were very cheap. In other cases (see Jacobsson, 1986, ch. 2) the use of CNC lathes can have a capital-saving effect even assuming one shift operation on account of this factor.

It is impossible to give a general answer to the question as to whether

NCMTs increase or decrease the amount of capital employed per unit of output. It seems as if it is partly a matter of the size of cost reductions for work in progress and for stocks. The other factor which needs to be taken into consideration, and which indeed appears to be decisive in a country like Sweden, is the institutional restriction to the full utilization of capital in the form of a lack of skilled workers who are willing to work shifts with conventional machine tools. In the Swedish context, NCMTs can clearly be capital-saving and this is to a large extent due to this factor. Indeed, in several firms we visited the main reason for choosing NCMTs was said to be the ability to find workers who accept shift work with the new machines. That NCMTs are frequently capital-saving in the Swedish context is corroborated by Boon (1984). Six out of eight firms interviewed by him claimed that machinery investment per unit of output was less for NCMTs than for conventional machine tools. In addition, practically speaking, all firms claimed that there was a reduction in stocks, work in progress and floor space.

Finally, there has been a significant reduction in the skills per worker needed to operate the new machine tools as well as in the number of people required to learn these skills. An operator of NCMT clearly needs some skills but the maximum time for an unskilled person, with a technically oriented secondary education, to be proficient with a NCMT was said by one Swedish firm to be 12 months at work, including in-house courses. Other firms suggested a maximum of 6 months. By contrast, for a qualified operator of a conventional lathe, five years' experience is often mentioned as being necessary to acquire proficiency. As far as the number of operators required to produce a given output is concerned, one firm which invested in CNC lathes needed to employ only 22 semi-skilled CNC lathe operators instead of 44 skilled operators. In another case, 21 CNC lathe operators substituted for 63 qualified operators. Hence for the operation of these lathes, the total *mass* of skills needed has been reduced in a very significant manner, not only because the average operator's skills have been reduced but because each operator produces a much greater output (Jacobsson, 1986, ch. 2).

The use of NCMTs, on the other hand, requires a set of new skills which are not important in the case of conventional lathes. First, in addition to the NCMT operators, setters and programmers are needed. About six to eight NCMTs are served by one setter and one programmer. These people are often former skilled workers who have joined the ranks of the white collar workers, or technicians.

Secondly, the repair and maintenance task has become more complex, mainly as regards the electronic part and interface with the mechanical parts. Given that for most users, apart from very large firms, the electronic part is of a 'black box' character, it is normal for the supplier to take over the repair and maintenance work. Usually the supplier of the numerical control unit provides the service on that unit while the machine tool supplier is responsible for the rest. It is clear that at the level of the machine, computerization has increased the skill requirements for repair and maintenance. However, as the number of machine tools is lower in the case of the use of NCMTs, the *amount* of repair and maintenance work needed to produce a given output has not necessarily increased (Senker et al., 1980).

If we translate the skill requirements into training years, we can calculate the *degree* of skill-saving with the use of NCMTs. Using the example in table 3.5, the total number of training years for the operation, setting and programming of the NCMTs would be 15 and in the case of the conventional lathes, 84. Hence the saving in skills amount to 82 per cent.[8] To this, however, we need to add the possibility that the greater skill content of the repair and maintenance staff of NCMTs reduces the extent of skill-saving associated with NCMTs. Furthermore, it has to be emphasized that certain skills, e.g. setters and repair skills, are necessary for the proper use of NCMTs. The number of individuals needing these skills is small, however, in relation to the number of skilled machine tool operators required with the use of conventional machine tools.

Of course, there is nothing inherent in the technique of NCMTs that dictates the use of semi-skilled instead of traditional skilled operators. A skilled worker can also operate a NCMT and will, in many cases, probably ensure a higher rate of utilization than a semi-skilled worker. However, the skill requirements *can* be reduced and the evidence suggests that on the whole the normal pattern is that semi-skilled workers *are* used (interviews with users of NCMTs, Rempp, 1982). Sometimes traditionally skilled workers are employed, but within a work organization that does not require the use of their full skills (Elsässer and Lindvall, 1984, ch. 8).

As we have indicated above, NCMTs can be greatly skill-saving.[9] In percentage terms per unit of output, the reduction in skill requirements is greater than that of labour, which in turn is far greater than that of capital. This means that the NCMTs represents a technical change, the appropriateness of which is a function of the degree of scarcity of skilled labour compared to semi-skilled labour and capital.

Notes

1   The 1978–82 data are based on official Swedish production and trade data. These data, however, are not satisfactory for several reasons. First, the production data appear to understate the value of production of NCMTs. This is easily demonstrated by a comparison with sales figures from individual firms. Secondly, for the trade data, only NC lathes, NC drilling and boring machines and NC milling and boring machines are identified. In the production data NC boring, grinding, polishing and lapping machines are included too. Thus, the official data probably underestimate the importance of NCMTs in total investment in

machine tools. In order to get a better idea of the diffusion of NCMTs we have therefore used also data from the Swedish Machine Tool Importers Association (SVAF). These data originate from the distributors of foreign made machine tools. Import over apparent consumption is very high in Sweden for machine tools, in the order of 80–85 per cent. In addition, we have used the official Swedish data for ascertaining the value of the Swedish producers' sales in the Swedish market as well as in its NC share. These data are an underestimate of the NC share but are used in the absence of other information. A complication here is that whilst the Swedish official data are based on production and trade data the data from SVAF are based upon orders. The SVAF data indicated that in 1983, 582 million Sw.kr. out of 1,100 million was for NCMTs. The official Swedish data for Swedish producer sales on the Swedish market were 217 million SEK out of which 143 million was in the form of NCMTs. Taken jointly, the figures suggest a NC share of 55 per cent. Had the data of SVAF only been used, it would have been 52.9 per cent. The same methods and sources were used to estimate the 1984 figure.

2 Due to the lack of data from other countries, the data on metal-forming machine tools come from the Federal Republic of Germany alone.

3 In 1978 firms with less than 100 employees had one-third of the 50,100 NCMTs installed in the USA or 16,000 NCMTs (Watanabe, 1983, p. 22). In 1983 there were 103,000 NCMTs installed in the USA: 40 per cent, or 41,000, of these were installed in firms with less than 100 employees. The market for NCMTs accounted for by these firms in the period 1978–83 amounted therefore to approximately 25,000 (41,000 − 16,000). The total market amounted to approximately 53,000 NCMTs: 25,000 out of 53,000 equals 47 per cent.

4 This figure is assumed to be half of the share of NCMTs that are installed in firms in the size group of 100–499 employees (see appendix table 3.4).

5 These paragraphs are based on Jacobsson, 1986, ch. 3.

6 This term is borrowed from Porter, 1980.

7 The Japanese industry captured very large shares of the world market as a result of implementing this strategy. For the case of CNC lathes, see Jacobsson (1986) for a detailed description and analysis of the structural changes in the period 1975–84.

8 In this calculation, we have assumed that the training time to reach proficiency is four years for operators of conventional machine tools and one year for operators of NCMTs. We have further assumed that setters and programmers have four years' training time. In this example seven CNC lathes are compared with 21 conventional lathes. Each has one operator. In addition, one setter and one programmer serve the NCMTs; 21 operators of conventional machine tools have in total 84 years' (21×4) training time; seven CNC lathe operators have in total seven years' (7×1) training time; the setter and programmer have jointly eight years' (2×4) training time. In total, the use of CNC lathes is associated with 15 years' training time $(7 + 8)$.

9 This saving of skilled labour per unit of output in the use of NCMTs does not, however, necessarily mean that the average level of skill requirements of all those involved in production, installation, programming, setting, optimization, operation, maintenance and repair of NCMTs decreases (see further section 8.2.1). This has to be investigated further.

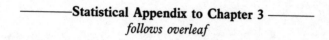

—————Statistical Appendix to Chapter 3 ————
follows overleaf

Appendix table 3.1 Share of NCMTs in total production of metal-cutting machine tools in a number of OECD countries,[a] 1976, 1982 and 1984, by type of machine tool

	1976 (US$m)	1976 (%)	1982 (US$m)	1982 (%)	1984[b] (US$m)	1984[b] (%)	Growth of production 1976–84 (%)
Boring machines							
NCMT	92	35	297	57	174	57	89
conventional	171	65	226	43	131	43	−23
total	263	100	523	100	305	100	169
Milling machines							
NCMT	145[c]	23	633	53	597	64	311
conventional	493	77	557	47	332	36	−33
total	638	100	1,190	100	929	100	46
Drilling machines							
NCMT	34	13	93	34	54	29	59
conventional	229	87	178	66	135	71	−41
total	263	100	271	100	189	100	−28
Gear-cutting machines							
NCMT	0	0	15	6	29	17	—
conventional	193	100	250	94	146	83	−24
total	193	100	265	100	175	100	−2
Grinding and polishing machines							
NCMT	10[d]	1	115	8	126	11	1,160
conventional	480	99	1,330	92	998	89	108
total	490	100	1,445	100	1,124	100	129

Lathes						
NCMT	479	30	1,403	61	1,492	73	211
conventional	1,112	70	885	39	559	27	−50
total	1,591	100	2,288	100	2,051	100	29
Other complete metal-cutting machine tools							
NCMT	441	30	1,617	38	2,084	62	372
conventional	1,016	70	2,639	62	1,236	38	22
total	1,457	100	4,256	100	3,320	100	128
of which machining centres	395	27	1,232	29	1,433	43	263
All metal-cutting machine tools							
NCMT	1,201	25	4,173	41	4,511	56	275
conventional	3,694	75	6,065	59	3,575	44	− 3
total	4,895	100	238	100	8,086	100	65

[a] USA, Japan, Federal Republic of Germany, France, Italy and the UK.

[b] For boring machines and lathes; excluding the UK.
For drilling machines, excluding the UK and Italy; US data are from 1983.
For grinding and polishing; excluding the UK and Italy.
For others; excluding Italy.

[c] For machining centres excluding UK. In the USA in 1976, in the case of NCMTs, 'all other metal-cutting and metal-forming machine tools', mainly NC milling, NC grinding and NC polishing machines were produced at the value of US$67.9 million. As very few grinding and polishing machines were being equipped with NC, we have assumed a US production of $60 million worth of NC milling machines (National Machine Tool Builders' Association, 1981/2, p. 101). As is shown above, grinding and polishing machines were still, in 1982, 92 per cent conventional.

[d] USA data are excluded as they do not show NC grinding machines separately.

Source: Elaboration on data supplied by CECIMO.

Appendix table 3.2 Share of NCMTs in the total production of metal-forming machine tools in the Federal Republic of Germany, 1976, 1982 and 1984, by type of machine tool

	1976 (US$m)	1976 (%)	1982 (US$m)	1982 (%)	1984 (US$m)	1984 (%)	Growth of production 1976–84 (%)
Punching and shearing machines							
NCMT	17	16	81	44	93	56	447
conventional	84	84	104	56	74	44	−12
total	101	100	185	100	167	100	65
Presses, including forging machines							
NCMT	5	1	28	6	22	7	340
conventional	376	99	420	94	300	93	−20
total	381	100	448	100	322	100	−15
Other complete metal-forming machine tools							
NCMT	2	0	15	3	43	13	2,050
conventional	394	100	427	97	300	87	−24
total	396	100	442	100	343	100	−13
All metal-forming machine tools							
NCMT	24	3	124	12	157	19	554
conventional	854	97	951	88	675	81	−21
total	878	100	1,075	100	832	100	− 5

Source: as in appendix table 3.1.

Appendix table 3.3   Sales of NCMTs to large and small enterprises in Japan, 1970–81 (in billion yen)

Year	Large firms	Medium and small firms[a]
1970	15,510	7,404
1971	17,278	8,639
1972	13,951	10,600
1973	23,075	25,122
1974	25,310	25,547
1975	12,921	17,756
1976	17,069	22,178
1977	23,820	24,856
1978	19,957	38,445
1979	49,013	79,017
1980	68,847	126,960
1981	92,068	153,292

[a] Firms with less than 300 employees.
Source: Jacobsson, 1986, ch. 3, based on Watanabe, 1983.

Appendix table 3.4 Distribution of NCMTs by size of firm in the USA in 1983

Class of firm[a]	No. of NCMTs	Share of total stock of NCMTs (%)
1–19	6,508	6
20–99	35,128	34
100–499	31,950	31
500+	29,722	29

[a] Classification by number of employees.
Source: Elaboration on *American Machinist*, 1983a, pp. 120, 121.

Appendix table 3.5 Average weight of CNC lathes made in Japan and the Federal Republic of Germany (in tons)

Year	Japan	FRG
1973	5.5	13.0
1974	5.8	12.9
1975	5.0	13.4
1976	4.9	13.7
1977	4.1	9.5
1978	4.0	7.4
1979	4.3	10.2
1980	4.6	9.2
1981	4.7	7.7
1982	4.8	8.2
1983	4.1	8.9
1984	4.2	8.4

Source: Jacobsson, 1986, ch. 3.

Appendix table 3.6 Production of CNC lathes in units by the
leading firms in Europe, the USA and Japan, 1975–82 (selected years)

	Top firm			*Average of following four firms*		
	1975	*1978*	*1981–2*	*1975*	*1978*	*1982*
Europe	n.a.	250	1,000	n.a.	210	590
USA	n.a.	n.a.	520[a]	n.a.	n.a.	n.a.[b]
Japan	270	1,000	2,500	105	525	1,400

[a] 1980.
[b] Total production of CNC lathes in the USA in units amounted to 2,739 in 1980, 2,021 in
1981 and 1,489 in 1982 (National Machine Tool Builders' Association, 1983/4, p. 100). As
the leading firm produces about 500 units, the next four firms must produce substantially
less per firm.
Source: Jacobsson, 1985, p. 13.

Appendix table 3.7 Production of machining centres in units by
leading firms in Japan, 1975, 1978 and 1982

	1975	*1978*	*1982*
Top firm	44	165	900
Average of following four firms	39	76	675

Source: Jacobsson, 1985, p. 13.

Appendix table 3.8 Basic data on diffusion of NCMTs in the USA in 1983

(1) Branches	(2) No. NCMTs	(3) % share in total stock of NCMTs	(4) % value added (US$m 1976)	(5) % share in total no. of machine tools	(6) (2)/(4)	(7) (3)/(5)
Fabricated metal products						
Cane and shipping containers	73	0.1	3,084	0.6	24	0.17
Cutlery, hand tools, etc.	982	0.9	4,486	2.5	208	0.36
Heating, plumbing fixtures	352	0.3	1,336	0.9	263	0.33
Fabricated structural metal products	4,278	4.1	10,048	5.5	426	0.74
Screw machine products	1,411	1.4	2,515	3.9	561	0.36
Metal stampings	1,866	1.8	7,554	4.4	247	0.41
Coating and engraving	580	0.5	1,788	0.5	324	1.00
Ordnance accessories	813	0.8	1,664	1.1	489	0.73
Miscellaneous fabricated metal products	4,158	4.0	6,670	4.3	623	0.93
Non-electrical machinery						
Engines and turbines	2,727	2.6	4,200	1.8	649	1.44
Farm, garden machinery	1,355	1.3	4,783	1.6	283	0.81
Construction, mining, material handling	7,641	7.4	9,946	3.7	768	2.00
Metal-working machinery	11,798	11.4	7,458	9.2	1,582	1.24
Special industrial machinery	4,037	3.9	5,174	3.7	780	1.05
General industrial machinery	5,815	5.6	8,043	5.0	723	1.12
Office, computing machinery	1,885	1.8	8,102	1.2	229	1.50
Refrigeration service machinery	1,321	1.3	5,214	1.5	253	0.87
Miscellaneous machinery excluding electrical	15,962	15.5	4,735	9.1	3,371	1.70

Electrical machinery					
Transformation, distribution equipment	520	0.5	2,702	0.9	192	0.56
Electrical industrial apparatus	1,775	1.7	4,916	1.9	361	0.89
Household appliances	722	0.7	4,454	1.1	162	0.64
Electrical lighting, wiring	1,011	1.0	4,204	1.9	240	0.53
Radio, TV equipment	185	0.2	2,544	0.4	73	0.50
Communication equipment	2,611	2.5	11,656	1.6	224	1.56
Electronic components	2,838	2.7	7,568	2.3	374	1.17
Miscellaneous electrical equipment	1,110	1.0	3,704	1.1	3.00	0.91
Transportation equipment					
Motor vehicles equipment	4,131	4.0	30,949	7.0	133	0.57
Aircraft and parts	8,782	8.5	12,735	4.5	690	1.89
Ships and boats	197	0.2	4,032	0.7	49	0.29
Railroad equipment	308	0.3	1,454	0.3	212	1.00
Motorcycles, bicycles, parts	187	0.2	—	0.3	—	0.67
Guided missiles, space vehicles	731	0.7	5,027	0.4	189	1.75
Miscellaneous transportation equipment	948	0.9	1,176	0.6	806	1.50
Precision instruments	4,874	4.7	16,386	5.0	292	0.94
Miscellaneous manufacturing industries	2,102	2.0	8,821	3.0	238	0.67
Total	103,308	100.0	219,112	100.0	471	1.00

Sources: Columns (1), (2), (3), (5), *American Machinist*, 1983a. Column (4), United States Department of Commerce, Bureau of Census, *Annual Survey of Manufacturers 1975–6.*

4

Industrial Robots

4.1 The Technique

In this study,[1] the following definition of industrial robots, which has been proposed by the International Organization for Standardization (ISO), will be used:

> The industrial robot is an automatic position-controlled reprogrammable multi-function manipulator having several degrees of freedom capable of handling materials, parts, tools, or specialized devices through variable programmed motions for the performance of a variety of tasks...It often has the appearance of one or several arms ending in a wrist. Its control unit uses a memorizing device and sometimes it can use sensing and adaptation appliances that take account of the environment and circumstances. These multi-purpose machines are generally designed to carry out repetitive functions and can be adapted to other functions without permanent alteration of the equipment. (ECE, 1985, p. 12.)

This definition is similar to the definitions used by the British Robot Association (BRA) and by the Robot Institute of America (RIA). However, the Japan Industrial Robot Association (JIRA) uses a much wider concept, including all of the following automation devices (which are classified by input formation and teaching method):

1	Manual manipulator	A manipulator that is directly controlled by an operator.
2	Sequence robot	A manipulator that functions by following a pre-established sequence.
	(a) fixed sequence	The pre-set sequence cannot easily be changed.

(b) variable sequence   The pre-set sequence can easily be modified.

3 Playback robot   A manipulator that can repeat any operation after being instructed by a man.

4 Numerically controlled robot   A manipulating robot that receives orders through numeric control.

5 Intelligent robot   A robot that can determine the functions required through its sensing and recognitive abilities.

On the basis of the ISO definition presented above, the 'manual manipulator' and 'fixed sequence' categories would not be classified as robots, but rather as automatic machines.[2] The manual manipulator and sequence robot categories (including variable sequence robots) are of the first robot generation. They have limited flexibility compared to second generation robots (i.e., playback and numerically controlled robots). The third generation robots (i.e., intelligent robots) differ from the previous generations by their sensory ability and capability to react to changes in their surrounding work environment.

4.2 The Diffusion of Industrial Robots

In figure 4.1 and appendix table 4.1 data on the stock of industrial robots are shown for certain OECD countries. Figure 4.1 shows that the growth in the stock of robots in the nine countries is remarkable. The average annual rate of increase in the number of robots installed in these countries was 44 per cent during the period 1974–84. However, annual cumulative growth rates differed considerably between countries. In Japan it was 52 per cent and in the USA it was only 27 per cent despite the fact that robots were first developed in the USA. The dominance of Japan over other OECD countries is striking. By the end of 1984 this one country accounted for 67 per cent of all robots installed in the countries included in appendix table 4.1. Between the end of 1984 and the end of 1985, the number of operating robots in Japan increased from about 65,000 to about 93,000 (or by 43 per cent) (Yonemoto, 1986), whilst the stocks in the USA, the Federal Republic of Germany and the UK increased by 7,000 (54 per cent), 2,200 (33 per cent) and 585 (22 per cent) units respectively during 1985 (Rooks, 1986, p. 3).

If we look at robot density, i.e. number of robots per 10,000 employees in the engineering industry, Japan is also the leading country, with 123 robots (see table 4.1). Sweden is second with 70; Belgium and Italy come next with slightly less than 30; the Federal Republic of Germany has 16 and the USA 15, i.e. only 12 per cent of Japan's density.[3]

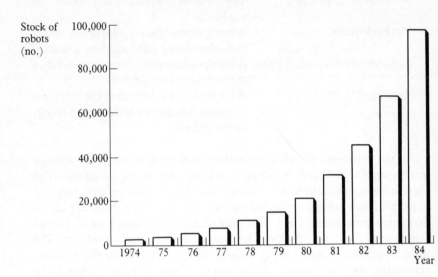

Figure 4.1 Total stock of industrial robots in nine OECD countries. We assume that the annual growth rate of the stock of robots for the years 1975, 1976, 1977 was the same as for the period 1974–8. The same applies for 1979, which is assumed to experience the same growth rate as for the period 1978–80.
(*Source*: appendix table 4.1.)

Table 4.1 Robot density in selected OECD countries

	No. of robots end of 1984	Employment in the engineering industry (ISIC 381-5) × 1,000		No. of robots per 10,000 engineering employees
Japan	64,657	5,275	(1980)	122.6
Sweden	1,900	271	(1980)	70.1
Belgium	860	306	(1980)	28.1
Italy	2,700	994	(1977)	27.2
FRG	6,600	4,082	(1980)	16.2
USA	13,000	8,809	(1980)	14.8
France	3,380	2,301	(1980)	14.7
UK	2,623	3,100	(1980)	8.5

Sources: Number of robots, appendix table 4.1. Employment, Japan – *Japan Statistical Yearbook*, 1983, p. 74; others – ILO, 1982, table 5B.

The information available to us on industrial robots in the Eastern European countries is fragmentary. We will mention a few figures to give some indication of the situation. In Hungary there were 17 robots installed in 1982 (ECE, 1985, p. 64). The robot population in the Soviet Union (including 'simplified' robots) was estimated at 6,000 units in 1980 (p. 51). In the German Democratic Republic, between 9,000 and 15,000 robots were in use in the beginning of the 1980s (manual manipulators and non-programmable fixed sequence robots are probably included in these figures) (p. 52). In March 1979 it was estimated that 720 robots were in use in Poland (p. 53).

Available information on the yearly investment in industrial robots for selected OECD countries is presented in appendix table 4.2. Japan is the largest investor, with US$ 800 million in 1984, followed by the USA and the Federal Republic of Germany. The global investment in 1984 was probably in the order of US$ 1,200–1,300 million and the total number of robots installed in that year was higher than 27,000. Of these, 65–70 per cent were installed in Japan.

However, the investment data in appendix table 4.2 include only the robots as such (and according to the ISO definition). If ancillary equipment and systems engineering are included, the investment figures should be increased by at least 70 per cent.

4.3 Applications of Industrial Robots

In terms of their applications or operational modes, industrial robots can be divided into three groups:

1 Handling robots: The workpiece is handled by the robot, for example, for material handling, loading and unloading of workpieces for machine tools, casting, pressing, injection moulding, forging, fitting etc.
2 Process robots: The tool is gripped by the robot, for example, in various types of metal-working operations (cutting, drilling, grinding, chipping), joining of materials (welding, gluing, wiring), surface treatment (paint spraying, surface coating, polishing) etc.
3 Assembly robots: Robots are used in the assembly of parts into components and complete products.

The application of robots with regard to these operational modes is shown in appendix table 4.3 for three countries. In 1980 process robots accounted for about half of the robot population in the UK and the Federal Republic of Germany but less than 30 per cent in the USA. Within this group spot welding was dominant in the USA and the FRG. Assembly robots were very few. Handling robots was by far the most important category in the USA, accounting for more than 70 per cent of the stock.

Over time, data are available for the UK and the FRG. Spot welding was still the dominant application in 1984 and process robots as a whole accounted for as much as 60 per cent in the FRG. The proportion of handling robots has decreased in both countries. The most interesting trend, however, is the growth in the proportion of assembly robots. The number of assembly robots increased from 246 to 452 units, i.e. by 84 per cent, during 1984 in the FRG (appendix table 4.3). During 1985, 301 assembly robots were installed in the FRG, i.e. the stock increased by 67 per cent to 753 units (Rooks, 1986, p. 3). During 1986 the number of assembly robots in the FRG increased by as much as 120 per cent (Lundström, 1987, p. 7). This certainly looks like the beginning of a breakthrough. The assembly robots category seems to be entering the growth phase of its S-curve.

A very large proportion of hours worked in the engineering industry is spent on assembly operations. In table 3.1 it was shown that assembly was the single most time-consuming task in the Swedish engineering sector in 1981; it accounted for 21 per cent of the total number of hours worked by blue collar workers. Hence, the possible breakthrough mentioned above would operate in a field with a large potential. The main obstacle to the widespread use of assembly robots has hitherto been of a technical nature. The assembly robots will have to 'learn to see and react' to their environment. Facilities like tactile sensors, optical sensors and versatile gripping devices are therefore essential for robot-based assembly. Considerable research and development efforts are currently being made in various industrialized countries to solve these problems. Clearly, in a number of cases, the technique is already so far developed that it can be usefully introduced in an industrial environment. One example is the introduction of an ASEA robot for the assembly of windscreen wipers at the Electrolux Factory at Ankarsrum in Sweden. The wipers were designed with robot assembly in mind. The main benefits are lower labour cost, increased possibilities for operation in several shifts and lower costs for work in progress (*Robot Application News*, 1985, p. 3).

4.4 Industrial Distribution of Robots

The engineering industry is the main user of industrial robots in all countries. In 1982, for example, this industry accounted for 88 per cent of the value of investment in industrial robots in Sweden (Carlsson, 1983, p. 43). In the case of Japan, this industry accounted for 70–80 per cent of the total robot purchase during 1976–82 (see table 4.2).

Table 4.2 shows that the automobile industry, along with the electrical

Table 4.2 Use of robots in Japan, 1976–82
(percentage share of total robot purchases, in value)

	1976	*1977*	*1978*	*1979*	*1980*	*1981*	*1982*
Industry							
Automobiles (incl. parts)	30	34	39	38	29	30	27
Electrical machinery (and electronics)	21	23	24	18	36	32	30
Metal products	6	3	7	8	5	5	5
Metal-working machinery	5	6	4	3	4	6	4
Other engineering products (boilers, engines, turbines, construction machines, ships, precision machines etc.)	7	8	6	7	4	4	5
Total engineering industry	69	74	80	74	78	77	71
Iron, steel and non-ferrous metals	8	7	5	6	4	4	2
Plastic moulding products	13	10	10	11	10	9	8
Other manufacturing	6	3	0	4	4	3	3
Non-industrial sectors	2	2	2	3	1	1	1
Exports	2	4	3	2	3	6	14

Manual manipulators and fixed sequence robots are included.
Source: Yonemoto, 1981, p. 240; Yonemoto, 1982, table 3; Yonemoto, 1983, figure 3.

machinery branch, has been the largest user of robots in Japan in value terms. The automobile industry is the main user of industrial robots in many OECD countries. In Canada, for example, it accounted for 63 per cent of the stock of robots installed in 1981 (OECD, 1983b, p. 42). This is partly a result of the historical development in robot technology, which was initially concentrated in spot welding and painting applications, both of which are important for the automobile industry. The automobile industry can be expected to remain a major user of industrial robots in the future. However, it could also be argued that by now this industry has made most of its investments in robots for welding and painting and that its relative share of robot investments will decline in the future, in spite of the fact that many of the robots used are gradually becoming obsolete and will have to be replaced. On the other hand, assembly robots will probably be used to a large extent for automobile parts production in the future.

Table 4.3 shows the distribution of robots within the engineering industry for the Federal Republic of Germany, The Netherlands, Sweden and the UK. In the UK and the FRG, transportation equipment (ISIC 384) accounts for more than half of the stock of robots. For Sweden one-third is in the automobile industry, and for The Netherlands only 6 per cent of robot stock is employed in the whole transport equipment branch, which reflects its relative smallness there.[4]

Table 4.3 Distribution of robots by sector within the engineering industry in terms of percentage share of the total stock (in number) of robots installed[a]

Branch	Sweden[b] 1983	Netherlands 1982	UK 1980	FRG 1981
Metal products (ISIC 381)	10	71	24	8
Non-electrical machinery (ISIC 382)	48	13	14	15
Electrical machinery (ISIC 383)	8	10	7	18
Transport equipment (ISIC 384)	34[c]	6	55	59

[a] Excluding the professional and scientific equipment branch (ISIC 385).
[b] The data on distribution in Sweden used here are very different from those presented for 1979 by the Computers and Electronics Commission, 1981, p. 199 and used in Edquist and Jacobsson 1984, p. 25 and UNCTAD TT/65, p. 34.
[c] The automobile industry only.
Sources: Sweden, Halbert, 1985b. UK, elaboration on OECD, 1983b, p. 41. FRG and The Netherlands, elaboration on OECD, 1983b, p. 42.

Finally, it should be noted that when compared with numerically controlled machine tools (NCMTs), robots have been used on work where flexibility is less important, e.g. on parts and products produced in larger batch sizes and in larger volumes with less variation and less reprogramming needs. This, however, is almost certainly a consequence of the still limited technical development of robots (UNCTAD, 1982, table 7).

Forty-five per cent of the stock of robots in Sweden in 1979 were found in the seven largest firms using them. Six of these were in the engineering industry: Ericsson, ASEA, Volvo, Sandvik, Saab–Scania and Electrolux. These firms also accounted for 25 per cent of the total NCMT stock and nearly 50 per cent of the total number of computer aided design-systems in the country (Carlsson and Selg, 1982, p. 88).

Outside the seven leading companies mentioned above, industrial robots were found in about 100 companies. This group was dominated by firms with only a few (1–5) robots. It may, therefore, be asumed that most robot-using firms in Sweden have developed a capability to use robots in only one or a few fields of application, such as welding or machine tool servicing (Computers and Electronics Commission, 1983, p. 20).

In Italy, Fiat had over 200 and Alfa Romeo about 40 of the 353 robots installed in the country in 1980 (*Industrial Robot*, 1981b, p. 176). Fiat, Alfa Romeo and Olivetti accounted for over 75 per cent of the robots in use (*Industrial Robot*, 1981a, p. 32). Hence, the concentration of robot users was even more pronounced in Italy than in Sweden. Of the 518 robots installed in Spain at the end of 1984 all but 16 per cent were in the automotive industry and Ford and General Motors' operations in Spain took up almost half of the Spanish robot population (*Industrial Robot*, 1985a, p. 32). In Belgium 69 per cent of the robot population are in the automotive industry (*Industrial Robot*, 1985a, p. 32). In all countries mentioned so far, large firms seem to account for a large majority of all robots installed. In the Federal Republic of Germany also large firms dominate and only one robot out of nine is installed in small and medium-sized firms (Utlandsrapport, 1984, p. 1).

In the UK, however, the stock of 371 robots existing in 1980 was spread rather thinly compared with the above-mentioned countries. There were between 140 and 180 robot-using companies (*Industrial Robot*, 1981a, p. 32). The number of multiple robot installations (more than 10) in one establishment was still very small in the UK in 1981. The vast majority of robot installations were of one or two robots (with the exception of the automobile industry). The BRA estimates that 80–100 British companies installed their first robots in 1981 (BRA, 1982b). This probably implies that robots are being introduced also by smaller firms, which means that a new large potential market for robots is being opened.

4.5 The Economics of Industrial Robots

According to a study of ten robot installations in Sweden (Computer and Electronics Commission, 1983), the dominant motive for investments in robots seems to be to decrease labour costs. Some companies also stress the objective of increasing the productivity of capital, through a decrease of the amount of capital tied up in raw material and components in the production process. Finally, many companies invest in robots in order to advance the firm technologically, i.e. robot investments are made for strategic reasons. In these cases, the firms do not care too much about the short-run profitability of the individual robot investment. Instead they consider it as an investment in learning which is expected to bring returns in the long run.

In five of the above-mentioned ten cases of robot installation, formal investment calculations were made *ex ante*. In two of these cases *ex post* calculations were also carried out. Hence, formal investment calculations were certainly not always the case, particularly when the investment was carried out for strategic reasons, e.g. to advance the company's technological capability. The most common calculation method used was the pay-off method. In general the profitability requirements for machine investments range between a pay-off period of one to five years, normally three. When the internal rate of return is calculated, the requirement is normally 20–30 per cent (Computers and Electronics Commission, 1983, pp. 36–7).

In order to illustrate the nature and magnitude of cost-saving from the introduction of industrial robots, we will refer to two examples of installations in Sweden. The examples are described by Carlsson (1983, pp. 27–35). Both *ex ante* and *ex post* calculations are presented.

The first example concerns an investment in an industrial robot for servicing a machine group consisting of two computer numerically controlled lathes, a measuring station and a conveyor. The pay-off time in this case was 2.68 years *ex ante* and 2.56 years *ex post*. The internal rate of return was 36 per cent *ex ante* and 38 per cent *ex post*. Most of the cost-saving was related to labour cost. (The labour saved through robot investments is in unskilled and semi-skilled labour.) The capital cost for work in progress also decreased, but on the whole the investment was labour-saving and capital-using.

The second example concerns an investment in two industrial robots for servicing a machine group consisting of two broaching machines and a drilling machine, connected by a conveyor. Here the number of workers needed decreased from 7.6 to 2.5 for the same operation. The *ex post* pay-off period was 2.4 years.

Another study of the profitability of industrial robots has been carried out in the Federal Republic of Germany (Volkholz, 1982). It showed that the pay-off period varied considerably between various applications and depended on the number of shifts per day. The shortest pay-off time was registered for loading and unloading of steel-shaping machines and of plastic injection-moulding machines as well as for welding applications. Somewhat longer pay-off periods were experienced for loading and unloading of workpieces for machine tools and for finishing mouldings. The pay-off periods for paint spraying and assembly processes were much longer.

If the pay-off period of an investment is less than 1.5 years, its profitability is generally considered very good. This was not the case in the above examples. According to well-informed sources (e.g. Computers and Electronics Commission, 1983), the general tendency is that the profitability of robot investments is somewhat lower than for average investments in machinery. This is not surprising since many robot investments are not realized except in the long term and are therefore not reflected in the usual investment calculations. It must also be stressed that the calculations made only referred to the robots as such. In order to assess significant effects on, for example, capital productivity, whole production systems should be studied. In other words, it is only when the robot is a part of a more comprehensive automation process that capital productivity (e.g. cost for work in progress) can be substantially improved.

In the short run and in a limited perspective, reduced costs for unskilled and semi-skilled labour is the most important reason for the majority of robot investments. However, when the firm automates various segments of the production process step by step, increased capital productivity may in the long run become a more important consequence. It is probable that it is not until the robot is integrated with other machines and the system as a whole is designed to attain rapid and flexible flows of components and parts that more positive economic or financial results can be attained.

Notes

1 An earlier version of this chapter was presented in Edquist and Jacobsson, 1986.
2 As we will see below, the broader definition used by JIRA leads to some non-comparability of statistics. However, attempts will be made to adjust the Japanese data, wherever possible.
3 We have chosen to use employment in the engineering industry as the denominator since most robots proper are used in this industrial sector. However, the picture is very much the same if workforce in all manufacturing is used. It could also be mentioned that until recently Sweden was considered to have a higher robot density than Japan. As shown in table 4.1, this was not the case in 1984. Moreover, it was incorrect also in 1981 and 1983, but correct in 1982. Using data from appendix table 4.1 and the same employment data as in table 4.1 the densities for 1980–3 were 26.5, 39.8, 41.4 and 88.6 for Japan. For Sweden they were 29.3, 35.1, 51.6 and 59.0 respectively. In spite of this, Sweden was listed as having a higher density than Japan in the September 1983 and the March 1984 issues of *Industrial Robot* (1983, p. 198; 1984, p. 41). the same is true for a brochure published by the Swedish Industrial Robot Association in August 1984 (p. 4). These incorrect statements are caused by the severe under-estimation of the Japanese robot population until the end of 1984, which is discussed in note (a) to appendix table 4.1. In the March 1985 issue of the *Industrial Robot*, however, the picture is corrected and Japan is listed as having a much higher robot density than Sweden (*Industrial Robot*, 1985a, p. 30).
4 In table 4.3 it may also be observed that metal products (ISIC 381) account for a very high proportion of the robots in The Netherlands. (It should also be noted that the total stock of robots in The Netherlands was only 77 units in 1982 – ECE, 1985, p. 65.) The same is true for non-electrical machinery (ISIC 382) in Sweden. Comparing tables 4.2 and 4.3 reveals an important difference between Japan on the one hand and Sweden, The Netherlands, the UK and the FRG on the other. The electrical machinery sector (including electronics) is a much more important robot user in Japan than in the other countries. (In comparing these two tables it must be kept in mind, however, that the Japanese figures include manual manipulators and fixed sequence robots. In addition, table 4.2 shows flow figures in value terms and table 4.3 shows stock figures in numbers.)

Statistical Appendix to Chapter 4

Appendix table 4.1 Stock of industrial robots in selected OECD countries (number of robots at end of the year)

	1974	1978	1980	1981	1982	1983	1984
Japan[a]	1,000	7,000	14,000	21,000	31,857	46,757	64,657
USA	1,200	2,500	3,500	5,000	6,250	8,000	13,000
FRG	130	450	1,133	2,300	3,500	4,800	6,600
Sweden[b]	85	415	795	950	1,400	1,600	1,900
Italy	90	300	353	450	700	1,800	2,700
UK	50	125	371	713	1,152	1,753	2,623
Belgium	n.a.	21	58	242	361	514	860
France	n.a.	n.a.	n.a.	n.a.	n.a.	1,500	3,380
Spain	n.a.	n.a.	n.a.	n.a.	n.a.	400	516
	2,555	10,896	20,615	31,140	45,470	67,374	96,737

Appendix table 4.1 continued

[a] Manual manipulators and fixed sequence robots are here excluded in the figures for Japan. It may be mentioned that until late 1984 the stock of robots proper in Japan was highly under-estimated in most sources outside that country. For example, the BRA, a source which is used very often, gives the figure 16,500 for Japan at the end of 1983 (BRA, 1983). For the end of 1982, BRA gives 13,000 units (BRA, 1982a). The Economic Commission for Europe uses the same figures (ECE, 1985, p. 42). In an earlier study we did not accept these figures but made our own estimate where we reached the figures 43,619 and 27,623 for 1983 and 1982 respectively (UNCTAD TT/65, p. 25 and Edquist and Jacobsson, 1984, p. 16). Hence, our estimate was better than the generally accepted – BRA – figures, but not entirely correct. In 1984 the BRA figures were dramatically increased from 16,500 for the end of 1983 to 64,600 for the end of 1984 (BRA, 1983: BRA, 1984). When exposed to our previous estimates mentioned above, the BRA responded by saying that BRA figures previous to 1984 were 'guestimates' but that in 1984 they received data from JIRA that fell into line with the BRA robot definition (Moutrey, 1985).

[b] The data on Sweden are estimates since a survey has never been done. We present here elaborations on estimates received from Halbert (1985a). According to Halbert (1985c), figures published earlier were too high.

Sources: 1974: Japan, Yonemoto, 1985; others, OECD, 1983b, p. 50. 1978: Belgium, British Robot Association (BRA), 1984; Italy, Industrial Robot, 1981b; Japan, Yonemoto, 1985; Sweden, Halbert, 1985b; others, OECD, 1983b, p. 50. 1980: Belgium, Denis, 1983, p. 197; Italy, Industrial Robot, 1981b, p. 176; Japan, Yonemoto, 1984, p. 152; Sweden, Halbert, 1985b; others, Cohen, 1983, p. 10. 1981: Belgium, Denis, 1983, p. 197; Japan, Yonemoto, 1984, p. 152; Sweden, Halbert, 1985b; others, BRA, 1981. 1982: Sweden, Halbert, 1985b; Belgium, Denis, 1983, p. 197; Japan, Industrial Robot, 1985a, p. 33 and Yonemoto, 1985; others, BRA, 1982a. 1983: Japan, Industrial Robot, 1985a, p. 33 and Yonemoto, 1985; Belgium, BRA, 1984; France and Spain, BRA, 1983; Sweden, Halbert, 1985b; others, Industrial Robot, 1984, p. 39. 1984: Japan, Industrial Robot, 1985a, p. 33, Sweden, Halbert, 1985b; others, BRA, 1984.

Appendix table 4.2 Yearly investment in industrial robots in selected OECD countries (number and million current US$)[a]

| | Sweden | | Japan[b] | | | | USA | |
| | | | All robots[c] | | Robots proper[d] | | | |
	Units	Value	Units	Value	Units	Value	Units	Value
Before 1976	575	13.7	16,400	165.0	n.a.	n.a.	n.a.	n.a.
1976	100	4.4	7,200	47.5	864[e]	20.5[e]	n.a.	18
1977	180	8.0	8,600	80.6	1,032	34.7	n.a.	26
1978	165	7.7	10,100	130.0	1,414	63.7	n.a.	30[f]
1979	269[g]	13.5[g]	14,500	193.6	2,755	92.9	n.a.	66
1980	n.a.	n.a.	19,900	345.4	4,577	224.5	n.a.	92
1981	n.a.	n.a.	22,100	487.8	8,177	356.1	2,100[h]	150
1982	215	10.3	24,800	596.0	14,880	506.6	n.a.	185
1983[i]	195	8.8	n.a.	n.a.	14,900	670.5	1,750	79
1984[i]	322	14.5	n.a.	n.a.	17,900	805.5	5,000	225

Appendix table 4.2 continued

	UK		FRG		Belgium		Italy	
	Units	Value	Units	Value[i]	Units	Value[i]	Units[j]	Value[i]
Before								
1976	n.a.	n.a.	n.a.	n.a.	n.a.	n.a.	55	1.3
1976	n.a.	n.a.	n.a.	n.a.	n.a.	n.a.	9	0.4
1977	n.a.	n.a.	n.a.	n.a.	12[k]	0.5	191	9.1
1978	45	n.a.	n.a.	n.a.	9	0.4	45	2.1
1979	} 246	n.a.	n.a.	n.a.	9	0.5	21	1.0
1980		10.0	n.a.	n.a.	28	1.4	113	5.5
1981	342	20.0[l]	1,167	68.3	184	10.8	237	13.9
1982	439	25.7[m]	1,200	57.5	119	5.7	452	21.6
1983[i]	601	27.0	1,300	58.5	153	6.9	1,100	49.5
1984[i]	870	39.1	1,800	81.0	346	15.6	900	40.5

[a] The exchange rates for the various years are taken from OECD, 1983a, p. 171. For the row 'Before 1976', the exchange rate used is that for 1975.
[b] Production figures are presented except for robots proper 1983 and 1984. However, the Japanese trade in robots has been very small until recently. Hence production data are a good proxy for investment.
[c] All the six categories normally classified as robots according to JIRA.
[d] 'Manual manipulators' and 'fixed sequence robots' are excluded. Percentage of all robots which are robots proper by value and number respectively for the 1977–82 period are as follows: 1977, 43, 12; 1978, 49, 14; 1979, 48, 19; 1980, 65, 23; 1981, 73, 37; 1982, 85, 60. (Source: Yonemoto, 1981, p. 239 for 1977–80; Kamata, 1983, p. 46 for 1981; Yonemoto, 1983, p. 23 for 1982.)
[e] It is assumed that the proportion between 'all robots' and 'robots proper' is the same as for 1977.
[f] Robots produced only for use within the producing company are excluded.
[g] The figure is a prognosis made by the Computers and Electronics Commission (1981).
[h] This is the only available figure for the USA with regard to the number of units installed. The cost per unit installed was $71,400 for 1981, which is considerably higher than for other countries (compare with figures cited in note (i)). The difference between the stock at the end of 1981 and 1980 in appendix table 4.1 is 1,500 units. This would give a unit cost which is even higher, i.e. $100,000. If the UK unit cost for 1981 ($58,500) is combined with the US investment value we would arrive at 2,564 robots installed in the USA during 1981.

Appendix table 4.2 notes continued

[i] The investment values are calculated under the assumption that the value of each robot is the same as in Sweden during the period before 1976, during 1976–9 and for 1982. It was assumed that the value is the same as in Japan for 1980 and as in the UK for 1981. For 1983 and 1984 we simply assume that the value of each robot was US$45,000, i.e. that the average value has recently been decreasing somewhat. This gives the following average values (US$) per robot: before 1976, 23,800; 1976, 44,400; 1977, 47,800; 1978, 46,700; 1979, 50,200; 1980, 49,100; 1981, 58,500; 1982, 47,900; 1983, 45,000; 1984, 45,000. The average for 1978–80 is then $48,700 and the average for the period prior to 1978 is $38,500. The unit values in this note have been used to calculate investment values for all countries with regard to 1983 and 1984 as well as for the FRG, Belgium and Italy.

[j] The total number of robots installed in Italy before 1980 according to appendix table 4.2 is 321 units (*Industrial Robot*, 1981b). According to the Italian Society for Industrial Robots (SIRI, 1983), this figure is 341 units. The two sourcs mentioned differ significantly with regard to 1980. The first one gives 32 units while the second gives 113 units.

[k] Total number installed prior to 1978.

[l] The figure given in the source is $40 million (=£20 million) invested in robot *systems* during 1981. It is assumed that half of this sum relates to robots, and the rest to ancillary equipment and systems engineering.

[m] Estimated under the assumption that the value of each robot is US$58,500, i.e. the same as in 1981.

Sources: Sweden: 1979 and before, Computers and Electronics Commission, 1981, p. 178; 1982, Carlsson, 1983, pp. 15–16; 1983 and 1984, Halbert 1985a. Japan: 1980 and before, Yonemoto, 1981, p. 238; 1981 and 1982, Yonemoto, 1983, pp. 18–19; 1983 and 1984, calculated as differences between stock figures in appendix table 4.1. USA: 1977, Livingstone, 1981, p. 65; 1976, 1979 and 1980, Cohen, 1983, p. 14; 1981, *Industrial Robot International*, 1982, p. 5; 1978 and 1982, Smith, 1983; 1983 and 1984, calculated from appendix table 4.1. UK: 1978–80, BRA, 1982a (units); 1980, interview with the BRA (value); 1981, BRA, 1982a, 1982–84, calculated from appendix table 4.1. Belgium: 1977–82, Denis, 1983, p. 197; 1983 and 1984, calculated from appendix table 4.1. Italy: until 1979, *Industrial Robot*, 1981b, p. 176; 1980–82, SIRI, 1983; 1983 and 1984, calculated from appendix table 4.1.

Appendix table 4.3 Robot applications in the UK, FRG and USA (stock of robots at the end of each year)

	UK 1980		UK 1983		UK 1984		FRG 1980		FRG 1983		FRG 1984		USA 1980	
Surface coating	69	18	167	10	177	7	155	14	586	12	727	11	290	5
Spot welding	59	16	349	20	471	19	339	30	1,560	33	1,894	29	1,190	19
Arc welding	48	13	234	13	341	14	138	12	856	18	1,334	20	270	4
Grinding, fettling, deburring, etc.	4	1	27	1	43	2	5	–	22	0	22	0	–	–
Process robots (total)	180	48	777	44	1,032	42	637	56	3,024	63	3,977	60	1,780	28
Assembly robots (total)	5	1	103	6	199	8	44	4	246	5	452	7	50	1
Die casting	33	9	38	2	40	2	56	5	132	3	147	2	880	14
Injection moulding	54	15	276	16	412	17	–	–	–	–	–	–	1,470	23
Machine tool servicing	39	10	165	9	213	9	113	10	320	7	466	7	–	–
Press tool servicing	21	6	48	3	59	2	28	2	121	2	135	2	–	–
Manipulation	8	3	–	–	–	–	34	3	–	–	–	–	120	2
Inspection, test	3	1	30	2	41	2	–	–	–	–	–	–	–	–
Palletizing, packaging	16	4	66	4	102	4	–	–	–	–	–	–	–	–
Handling, general	2	–	22	1	24	1	170	15	775	16	920	14	1,950	31
Handling robots (total)	176	48	645	37	891	37	401	35	1,348	28	1,668	25	4,420	71
Research, education	–	–	81	5	122	5	17	1	80	2	157	2	–	–
Others	10	3	147	8	188	8	34	3	100	2	346	5	–	–
Total	371	100	1,753	100	2,432	100	1,133	100	4,798	100	6,600	99	6,250	99

Corresponding data for UK and FRG 1981 and 1982 are excluded here but available in Edquist and Jacobsson, 1984, p. 21 as well as in UNCTAD TT/65, p. 22.

Sources: BRA, 1981; BRA, 1982a; BRA, 1983; Cohen, 1983, p. 63; OECD, 1983b, p. 37; BRA, 1984.

5

Flexible Manufacturing Systems

5.1 The Technique

The diffusion of individual, or stand-alone numerically controlled machine tools (NCMTs), has now become rapid, as shown in chapter 3. The main development in the past decade, that is, since the introduction of microcomputer-based numerical control units, has been the standardization and mass production of stand-alone NCMTs. In parallel, important developments have taken place within the fields of robotics, measuring and ancillary equipment. The next step in the process of automating the engineering industry is the building of production *systems* which are both flexible and automated. As an overall term for these types of systems 'flexible manufacturing systems' (FMSs) is often used although this term covers systems of different sizes and with different degrees of automation.[1] In this book, we will use the terms as specified below.

1 A *flexible manufacturing module* (FMM) consists of: a stand-alone NCMT; material handling equipment, such as a robot or a pallet changer; as well as some kind of a monitoring system. The FMM can be incorporated as a module in a larger system (*Metalworking, Engineering and Marketing*, January 1984, p. 49).

2 A *flexible manufacturing cell* (FMC) consists of at least two conventional and/or NCMTs. Like the FMM, the FMC includes a material handling device. This could be a robot serving a number of machine tools standing in a circle or line, or automatic pallet changers in conjunction with automatic transport between the NCMTs, e.g. some kind of conveyor (*Metalworking, Engineering and Marketing*, January 1984, p. 49).

3 A *flexible manufacturing system* (FMS) proper contains, in Dr Warnecke's words (1983, p. 682), 'several automated machine tools of the universal or special type and/or flexible manufacturing cells and, if necessary, further manual or automated workstations. These are interlinked by an automatic

workpiece flow system in a way which enables the simultaneous machining of different workpieces which pass through the system along *different routes*'.

4 A *flexible transfer line* contains, again in Dr Warnecke's words (1983, p. 682), 'several automated universal or special purpose machine tools and further automated workstations as necessary, interlinked by an automated workpiece flow system according to the line principle. A flexible transfer line is capable of simultaneously or sequentially machining different workpieces which run through the system along the *same path*'.

5 A *fixed transfer line* consists of a number of special purpose machine tools which are designed to produce one product only. After a very long setting time, a fixed transfer line can produce a different variant of the product, e.g. a different size of cylinderheads.

The choice of level of automation is normally a function of the degree of flexibility required. In figure 5.1 we can see how the choice of level of automation is a function of the number of different types of workpieces to be machined and the yearly production of each variant. When many variants are produced in very small annual amounts it is still most economical to choose stand-alone NCMTs or even conventional machine tools. As we progress to larger annual production of fewer variants, the more complex

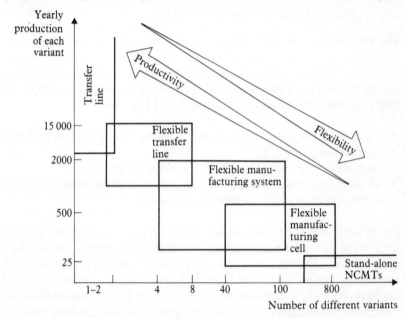

Figure 5.1 Choice between different manufacturing concepts. FMMs are not included in the figure but should cover the western part of the NCMT area and the eastern part of the FMC area. (From Spur and Mertins, 1981.)

will the systems become. Fixed transfer lines are normally used when there are only one or two variants which are mass produced, whilst flexible transfer lines are used when the number of workpieces of different kinds range between 2 and 8. In figure 5.1, FMSs are chosen when the number of variants is between roughly four and 100. In a review of 57 FMSs, Spur and Mertins (1981) showed that in 72 per cent of the cases, the number of different parts produced was less than 100 but in as many as 44 per cent the number of different parts was less than 10. In a slightly larger survey, (83 FMSs) we found that 77 per cent of the FMSs less than 20. Finally, ECE (1986) reports from a study of 157 FMSs that 33 per cent produced less than ten variants and 69 per cent less than 100 variants. Hence, a substantial part of present FMSs are relatively inflexible.[2] However, it should also be noted that almost 30 per cent of the FMSs surveyed by Spur and Mertins (1981) produced more than 100 different parts, thus showing that they were highly flexible. The same is true for the ECE (1986) study where 31 per cent of the systems produced more than 100 variants.

5.2   The Diffusion of Flexible Manufacturing Systems

5.2.1   FLEXIBLE MANUFACTURING MODULES

The first step toward system development is the automation of the loading and unloading of parts to and from a NCMT. That is, functions (2) and (7) of those listed in section 3.1 above (see pp. 23–4). Based on, for example, a computerized numerically controlled (CNC) lathe, functions such as the following must be included (Nordström, 1983, p. 169).

- Flexible, easily settable, integrated handling equipment, for different configuration of parts.
- Automatic tool change system – functions (3) and (6).
- Monitoring systems like tool breakage control, programmable wear time, automatic measuring, diagnostics – function (9).

For lathes, a robot (a conventional multipurpose one or gantry type[3] is used as the automatic parts changer. For machining centres, the development of the automatic pallet changer provided the basis for unattended machining (Bryce, 1983). For between 50 and 150 per cent of the cost of the basic NCMT, a considerable increase in the level of automation can take place. For example, a CNC lathe equipped with a special parts handling unit can operate unmanned for up to six hours. This means that a third unmanned shift can be introduced, in addition to increasing labour productivity dramatically in the first and second shifts. As will be shown in chapter 7, capital productivity may also increase, although this is not always the case.

The diffusion of such units is very recent. As Ashburn (1981) noted

> The trickle of untended machines offered in Japan became a flood at the 4. EMO [Exposition Européene de la Machine Ontil], the recent international show in Hanover. There were more machines with robots and loaders (nearly 200) than one would have found with any form of NC only a few years ago.

There is, however, a long lag between the display of prototypes at exhibitions and a significant diffusion in industry. In January 1983 the leading Japanese firms claimed that only 10 per cent of their production of CNC lathes were equipped with programmable devices for the material handling function.[4] Similarly, the leading Swedish producer of CNC lathes sold more than 70 CNC lathes with a programmable automatic material handling device between the end of 1979 and mid 1983 which is roughly equal to 10 per cent of CNC lathes produced. Thus, even for this, the most simple form of system development, diffusion was still in its introductory phase until very recently.

The rate of diffusion of FMMs appears, however, to have increased in the past two years. For example, of the 246 CNC lathes sold in Sweden in the period September 1984 to August 1985, 56 (23 per cent), were installed with a dedicated material handling device (excluding those pieces of dedicated equipment which feed automatic and CNC lathes with steelbars). In addition, an estimated[5] 50 CNC lathes were installed in conjunction with stand-alone robots that dealt with material handling (see further section 5.5). In total then, over 40 per cent of the CNC lathes sold in that period were equipped with an automatic material handling unit.

5.2.2 FLEXIBLE MANUFACTURING CELLS

The basic component of a FMC are: a number of machine tools (two to five machine tools can be seen as being the rule), not necessarily NCMTs, and a material handling unit (frequently a robot or two). Often, the information flow to and from the FMC is integrated into a larger control system (DEK, 1983, pp. 27, 25).

FMCs are less flexible than stand-alone NCMTs or FMMs. Normally, FMCs are applied to a family of components, e.g. shafts within a given size range. Thus, the FMC is designed to produce a narrower range of components than the above mentioned units. Within this range, however, the FMC is very flexible.

Each FMC has to be designed to suit the requirement of each customer, although standard modules can often be used. There is therefore a great flow of information and a lot of consultation between the FMC supplier and the user, which is a factor contributing to the demand's being limited so far mainly to large firms. Another factor leading to the predominance of large firms using FMC (DEK, 1983, p. 27; Warnecke, 1983) is the risks involved in buying a FMC. It is technically complicated to link different machines and product flows and this implies that the costs of the system often become higher than anticipated. Furthermore, there is a risk that it may not function very well, at least initially. It may take several years before the system begins to pay off. The technology is, however, becoming more reliable and a future diffusion on a large scale is anticipated (DEK, 1983, p. 27). At the moment, however, there is a relatively limited supply of these FMCs and several specialist producers often have to team up to supply the system and to solve the problems of interfacing.

There are no exact data on the number of FMCs installed in the OECD economies. A study of the data from various countries given below would, however, indicate that the technique is still in its introductory stage:[6]

1 In 1984, only 7 per cent and 9 per cent of the installed robots in the Federal Republic of Germany and the UK, respectively, served machine tools (see appendix table 4.3). Indeed the total number of robots serving machine tools in the FRG was only 466 in 1984 whilst in the same year there was an estimated total stock of 46,000 NCMTs (see table 7.2). In the UK, the total stock of robots serving machine tools was only 213 units in 1984 whilst there was a stock of 32,500 NCMTs (see table 7.2).

2 One of the largest robot producers in the world, ASEA of Sweden, have stated that only 15 per cent of their robots are applied to machine tool cells (Helliwell, 1983), that is, roughly 150 robots per year.[7] All in all, ASEA had sold 500 robots serving machine tools (excluding presses) by mid 1985 (Rizwi, 1985).

3 ECE (1986) reports that there were 525 FMCs in the USA by the end of 1984. By comparison, 103,000 NCMTs had been installed by 1983.

5.2.3 FLEXIBLE MANUFACTURING SYSTEMS[8]

FMSs can be said to have the following characteristics:

• The flow of tools and parts between different machine groups is automatic: using trucks, conveyors, cranes etc. Normally, the workpiece handling is *not* carried out by industrial robots, instead automatic pallet changers and custom designed equipment are used. The relatively rare use of robots can partly be explained by the fact that only few FMSs incorporate CNC lathes. Robots are often used on material handling for CNC lathes that machine rotational parts, whereas most FMSs consist of machining centres for the production of prismatic parts, e.g. gearbox housing.

- Very low utilization of labour.
- A high flexibility (in relation to fixed transfer lines) in that the system can quickly be switched to produce different products. The flexibility of flexible transfer lines is less than for flexible manufacturing systems. Warnecke (1983) distinguishes between short- and long-term flexibility. A FMS with long-term flexibility has the ability to adapt to unforeseen changes in the manufacturing task. Today, however, the FMS, as the FMC, is normally designed very carefully to accommodate only a family of parts, i.e. they are only flexible in the short term and to a limited degree.

FMS systems are at a very early stage of their diffusion. Steinhilper (1985) suggests that in 1984 there were 230 FMSs installed world-wide. Of these, Japan accounted for 100, the USA for 60, the FRG for 25 and Sweden for 15. ECE (1986) suggests that there were around 350 FMSs in 1984/85. As a comparison, Hatvany et al. (1981) estimated that there were 120 FMSs installed world-wide in 1980. Another way of demonstrating the early stage of diffusion of FMSs would be to compare the number of NCMTs that are incorporated in FMSs with the total stock of NCMTs. ECE (1986) suggests that 2,139 machine tools, not all NCMTs, were incorporated in 309 FMSs studied in 1984/85. As far as the stock of NCMTs is concerned, we have stated that in the USA alone there were 103,000 NCMTs installed by 1983; in the FRG there were around 46,000 NCMTs in 1984 and in the UK there were 32,500 in 1984 (Machine Tool Trades Association and Metalworking Production, 1985). Thus, an insignificant number – less than 1 per cent – of the world's NCMTs are incorporated into FMSs. In terms of flow, ECE (1986) estimates that 2–3 per cent of the world machine tool market was made up by FMSs by 1985.

5.3 The Industrial Distribution of Flexible Manufacturing Systems

Where then, in terms of industries, have FMSs been applied? In other words, which types of products are and will be feeling the earliest impact of FMSs?

In terms of industrial sectors, we undertook a survey which produced information on 129 FMSs. The distribution, by sector, is shown in table 5.1. As compared with the pattern of diffusion of NCMTs, we may note that the share of 'non-electrical machinery' is roughly the same, whilst that of transportation equipment is far larger for FMSs than for NCMTs. The reverse is true for electrical machinery and metal products.

Table 5.1 Distribution of 129 FMSs by sector

Sector	No.	%
Metal products (ISIC 381)	6	5
Non-electrical machinery (382)	67	52
Electrical machinery (383)	7	5
Transportation equipment (384)	46	36
Precision instrument (385)	3	2
	129	100

Source: The data base for this, and subsequent tables, was assembled from the following sources: *American Machinist*, 1983b; A:son-Stråberg, 1983; Dupont-Gatelamond, 1983; Goebel, 1983; Hatvany et al. 1981; Helliwell, 1983; Ingersoll Engineers, 1982; IFS Publications, 1983; McBean, 1983; Mertens, 1981; *Metalworking, Engineering and Marketing*, January 1982, May 1982, July 1982, November 1982, January 1983, May 1983, July 1983, September 1983, January 1984; Mueller, 1983; Popplewell and Schmoll, 1983; Purdon, 1983; Spur and Mertins, 1981; Wiehe, 1983.

The relatively large proportion of FMSs in the automobile industry can largely be explained by the fact tht FMSs are often applied to large volume production. Indeed, out of 127 FMSs for which we have information, the parts produced in the FMSs are incorporated into automobiles and trucks in 27 cases, or 21 per cent. Other important final products incorporating parts manufactured by FMSs include machine tools, construction machinery, tractors and aerospace components. Jointly, these final products account for 60 per cent of installed FMSs. Other important final products are electric motors, diesel engines, pumps, valves and compressors (see appendix table 5.1).

Amongst the firms that have installed these FMSs, particularly notable are Caterpillar (9), Yanmer diesel engines (5), Volvo (4), Sumitomo (4), Ahlstrom Atlantique (3), Hitachi (3), Ford (2), International Harvester (2), General Motors (2), IBM (2). These are all very large firms. According to one source (Rempp, 1982), almost all FMSs installed in Japan and the USA are found in larger companies whilst in the Federal Republic of Germany, a third of the FMSs are installed in somewhat smaller plants with 500–1,600 employees. According to another source, most of the users of FMSs in the world are multinational corporations (Ingersoll Engineers, 1982, p. 50). Large firms are mainly responsible for introducing FMSs for the same reasons as for FMCs: a need for in-house expertise and a high risk of failure, as well as high investment costs and, in the case of FMSs, the need for very large outputs to justify investment.

In the FMS, production is restricted to parts or sometimes complete components. In a sample of 103 FMSs, different types of housing, mainly but not only for automobiles/trucks/tractors and construction machinery, accounted for 45 per cent of the FMSs. Shafts, gear parts and cylinderheads account for another 27 per cent (see appendix table 5.2). These products are typically produced in fairly large volumes but with some variations in size and with several areas of application. For example, John Deere produces gearboxes for tractors and construction machinery in eight different variants and with a yearly volume of altogether 50,000 units (A:son Stråberg, 1983). On the other hand, very few FMSs produce parts for unique areas of application, in particular FMSs are *not* used for such products as special industrial machinery, with the exception of machine tools. The extensive use of FMSs in this industry can partly be explained by the fact that these firms are the main suppliers of FMSs. In other machinery industries, the production volumes are probably too low to justify building a FMS. For example, special industrial machinery is normally produced in batches of a few hundred units or less, and Volvo suggests that the minimum efficient batch size for the use of FMSs in the production of engine blocks is 2,000 units (*American Machinist*, 1983b).

5.4 The Economics of Flexible Manufacturing Systems

Flexible manufacturing systems, including FMMs and FMCs, are introduced to meet four general objectives:

1 Improved machine utilization. One important objective of FMSs is to improve utilization. As Hull (1983, p. 538) puts it, 'of the 8,760 hours in a year available, probably 6,000 or more will pass in the typical machine shop with plant standing idle and unattended'.
2 Reduced costs for work in progress. Long manufacturing lead times results in high rates of interest to be paid and/or loss of interest earned. In a conventional machine shop lead times are often very long, individual batches can take six to nine months to complete the return trip to the stores (Hull, 1983, p. 539). Moreover, a depressing condition applies to the workpiece as it travels around the plant. Less than 5 per cent of its manufacturing lead time is spent in receiving any attention, and during less than 2 per cent of the time is value being added. Another important objective of flexible automation is thus to reduce the cost of work in progress (Hull, 1983, p. 538).
3 Increased labour productivity. In spite of the introduction of NCMTs,

which has brought about increased labour productivity, the machinery process is nevertheless a fairly labour-intensive activity.

4 Greater flexibility of the equipment as compared to fixed automation machinery. Faster reaction to changing market demands is emphasized as a competitive tool by firms. The increased flexibility of the production equipment also implies that the amount and cost of stocks can be reduced.

One needs, to note, however, that there are a number of ways of achieving these objectives, of which only one is applying new hardware. First, a major factor restricting improved machine utilization in conventional workshops is not technical but social. Shift work is not popular, for obvious reasons, and in particular it is not appreciated by skilled machine tool operators.

Secondly, in order to reduce the costs for work in progress and stocks, new organizational principles can be applied to conventional technology, be it either manual machine tools or stand-alone NCMTs. A faster through-put time and, therefore, an ability to react more quickly to market demand as well as to reduce stocks, can be achieved by applying the concept of group technology. Briefly, group technology implies establishing families of products, e.g shafts of different sizes, and organizing a production line in which all the production processes that are required to complete the product are present. Such an organization is distinct from the traditional layout principles where all machine tools of a specific type, e.g. lathes, are grouped together and the workpiece is transported between different groups of machine tools all around the factory.

Group technology can be applied with conventional machine tools but is often associated with the introduction of NCMTs. These machine tools are then arranged in, for example, a circle. The workpiece is transported between the machines either by hand, by conveyor or other automatic means. The loading and unloading can also be done manually or automatically. To the extent that the transport of the workpiece and the loading/unloading is done automatically we have a FMC or a FMS. One particular aspect of group technology is that the overhead costs can be reduced. Indeed, one observer deems this to be the main source from introducing a FMS. As Dempsey (1983, p. 5) explains:

> By closely linking operations, the FMS is bringing under control all the overhead costs which occur between operations such as accumulating batches of parts, rejected or damaged components, others getting lost or components leaving the shop for special operations only to disappear for months or any of the other things that occur which together make up the sources of delay and overhead costs in a factory. Correctly applied, the FMS principle does away with mountains of paperwork customarily associated with these obsolete factories, and

the droves of people who handle the paperwork or spend their time putting right things that should never have gone wrong in the first place.

Thirdly, the advantages of having automated the transport and workpiece loading and unloading functions are mainly that the labour input is reduced and that social restrictions to improved machine utilization in the form of introducing several shifts are overcome.

Evidence on the economic efficiency of stand-alone NCMTs, FMMs, FMCs and FMSs varies in availability. The economic efficiency of NCMTs (stand-alone) is well established and was discussed in chapter 3.

As compared with stand-alone NCMTs, one advantage with FMMs is increased labour productivity as the machining can continue untended for some hours. Another, related, advantage is that a third, night, shift can be introduced which increases the machine (and building) utilization. Furthermore, the utilization time (the amount of actual machining time in total available time) can increase due to the fact that less time is needed for fixing the workpiece to the worktable. For machining centres this is done outside of the machine in special stations. The reduced setting time means greater flexibility and therefore stock levels can be reduced. The machine can also work on through employees' coffee breaks or lunch hours and does not need to stop when personnel is changed from one shift to another. Obviously this improves the machine utilization.

A FMC is an automated version of the group technology principle of organization. As compared to the FMM, the FMC can further reduce the costs for work in progress and stocks of final products as the workpiece is machined completely within the group. The Boston Consulting Group (1985, p. 12) suggests that paybacks for FMCs usually range from two to four years with a return on investment of 2–5 per cent over the rate of inflation.

Concerning the economics of FMSs, let us consider the words of five authorities in this field: In a study in 1981 of FMSs within the Federal Republic of Germany, Rempp notes:

> The companies were unsure about the economic efficiency of FMS. The inquiry showed that the efficiency of the installed FMS is much lower than assumed in many publications until now. Also a lot of planned FMS have been cancelled as a result of feasibility studies... the causes for lack of efficiency are generally in the transportation link-up system. (1982, p. 13.)

Rempp further adds that

> ...efficiency calculations are difficult to carry out with any precision. Most of the systems installed so far in West Germany do not meet the requirements of a quantified economic feasibility test. (p. 20.)

Dempsey notes:

> However, there is a fundamental and confusing problem emerging
> with all this FMS activity. It is that, in general, although FMS are
> known to give us 20–50% reduction in unit cost of a component or
> sub-assembly, these benefits are not being felt on the all important
> financial operating statement. This view is supported by the interviews
> in a recent survey of more than 50 installed FMS in Europe, USA and
> Japan which indicated that only two or three appeared to provide
> anything like the actual financial performance predicted for them.
> (1983, pp. 3, 4.)

Kobayashi notes:

> The various FMSs observed at the Chicago and Osaka [exhibitions]
> sites have one point in common: they are all nearing practical applica-
> tion. Technically speaking, for example with regard to the linking
> of machine tools and transfer system, the computers for centralized
> monitoring and control use and various types of build-in sensors, the
> systems have reached the stage of practical use...However, it should
> be pointed out that many of the FMSs we have today are still in their
> infancy. Problems still remain at the economic level, which is obviously
> the prime concern of the customer...At both shows, there were very
> few builders who, when approached, could give satisfactory answers
> regarding payout periods, either positive calculations or estimated
> period. (1983, p. 15.)

The Swedish Computers and Electronics Commission notes:

> FMSs must to a large extent be made on a custom designed basis.
> This does not, however, mean that there are not good possibilities
> for developing general concepts and system modules for different
> application areas and system configurations. System modules are
> necessary if it is going to be possible to reduce the development costs
> for the individual users. (DEK, 1983, p. 30, our translation.)

The Boston Consulting Group suggests that

> ...there is now growing evidence that most FMS installed in the late
> 70s and early 80s are plagued by technical problems which have led
> to very low rates of capacity utilization and often negative financial
> returns. In a number of cases, this had been expected, the systems
> having been considered as R&D investments. (1985, p. 12.)

The Boston Consulting Group also claims that

> In Japan, end users are becoming increasingly aware that flexible cells provide them with most of the benefits of automation (such as an increase in cutting ratio, and a reduction in lead times), without the penalties of extremely high investments and technical problems. The cell market is already growing rapidly. (p. 13.)

Hence, in spite of the fact that FMSs mean considerable manpower savings in comparison with conventional technology (ECE, 1986), the economics of FMSs are still to be proved in a convincing manner. The lack of competitiveness stems partly from the need to custom design the system[9] and partly from a lack of development in the transportation and information systems. Up until recently at least, the majority of systems were experimental. Another factor is the fact that there are only a handful of firms in the world that supply complete FMSs.

All these factors, however, are in the process of change. ECE (1986) suggests that FMMs are becoming economically a more attractive option than they formerly were and cites one study which claims that 'typical' cost reductions of FMS compared with conventional systems amount to between 14 and 27 per cent. Larson (1985) cites another study which claims that the average savings in production costs in 13 European FMSs studied amounted to 25 per cent. In addition to reduced costs of production, both studies, as well as Bessant and Hayward (1986), indicate that reduced lead time is a major benefit for the user of an FMS. We would therefore expect a faster rate of diffusion of FMS, in the next decade. However, even a representative of one of the few suppliers of FMSs, Cincinatti Milacron, suggested in 1983 that

> Despite the present high intensity of interest and effort in FMS it is likely to be a slowly maturing field. This is not just based on an extrapolation of the past, but because each system is still to some extent a research platform for the next step, and the cycle time of each advance is 2 years. (Hull, 1983, p. 544.)

5.5 The Case of Sweden

In order to be able to provide a clearer picture of the trends towards 'system development' in machining, we made a special study of Sweden. Sweden is at the forefront in the application of new techniques (see, for example, table 7.2) and we expect that the pattern revealed in our study will be relevant for other leading industrial nations in the years to come.

In this section we will first present our estimates of the stock and flow of NCMTs, FMMs, FMCs and FMSs in Sweden. A comparison with some other OECD countries will subsequently be made. The reader is referred to the appendix to section 5.5 for a detailed discussion of the estimation procedure and our sources.

In table 5.2 we have set out key data on the stock and flow of NCMTs, FMMs, FMCs and FMSs. In addition to the number of respective techniques, we have also indicated the estimated number of NCMTs that are incorporated in the various types of systems. Finally, the share of these NCMTs in the total stock and flow of all NCMTs is also presented.

In terms of *stock* figures, there were approximately 6,000 NCMTs, 200 FMCs and 15–20 FMSs in 1985. There are no available data on the stock of FMMs. Excluding FMMs, in the order of 500–600 of the NCMTs were part of a system. This amounts to between 8 and 10 per cent of the stock of NCMTs.

In terms of *flow* figures, approximately 700 NCMTs were sold in the period September 1984 to August 1985. Rather more than 200 of these were estimated to be incorporated into some kind of system. We estimate that at least 80 FMMs were sold as well as approximately 50 FMCs and three FMSs. Jointly, the NCMTs estimated to have been sold as a part of a system amounted to approximately 30 per cent of the total flow of NCMTs.

Several observations can be made on the basis of these data. First, the number of NCMTs incorporated into a system is substantially greater for the flow figures than for the stock figures. This is especially so for FMSs, but we should note that the flow figure for FMSs is probably an under-estimation as the data exclude not only such FMSs as are put together in-house by the user, see the case of Volvo Components in section 7.2.2, but also metal-forming FMSs. Secondly, a very large part of the flow of NCMTs, around 30 per cent, is included in some form of a system. Thirdly, the diffusion of systems takes place mainly in the form of FMMs and FMCs.

The diffusion of systems in Sweden is thus most significant. It is not possible to make a really satisfying comparison with other countries since the equivalent data for other countries are not available to us. A preliminary comparison can be made, however, on the basis of three sets of data.

Firstly, as is shown in table 7.2, the density in use of FMS is much greater in Sweden than in other developed countries. Secondly, we can estimate the diffusion of FMCs by the rate of installation of robots serving machine tools. Data are available for the UK and for the Federal Republic of Germany.

In the UK the number of robots sold in 1984 to serve machine tools amounted to 48 units. In the same year, 3,897 NCMTs were sold. Making the assumption that one robot serves 2–2.5 NCMTs, a total of 96–120 NCMTs were incorporated into FMCs where robots carry out material

Table 5.2 Key data on the estimated stock and flow of NCMTs, FMMs, FMCs and FMSs in Sweden

	Stock[a]			Flow[a]		
	(1)	(2)	(3)	(4)	(5)	(6)
	Units	Total no. NCMTs incorporated in system	Share NCMTs in (2) in total stock NCMTs (%)	Units	Total no. NCMTs incorporated in system	Share NCMTs in (5) in total flow NCMTs (%)
Technique						
NCMT	6,000	–		700	–	–
FMM	n.a.	n.a.	n.a.	80[b]	80	11
FMC	200	400–500	6.7–8.3	50	100–125	14–18
FMS	15–20	90–120	1.5–2.0	3[c]	17[b,c]	2

[a] Flow refers to September 1984 to August 1985 and stock refers to 1985.
[b] Only those including CNC lathes and machining centres.
[c] Excluding those made in-house by the user on the basis of already existing NCMTs.
Source: See appendix to section 5.5.

handling. This amounts to at most 3.1 per cent of the NCMTs sold in 1984. In the case of the FRG, 146 robots serving machine tools were sold in 1984. The flow of NCMTs amounted to 8,679 in the same year (VDMA, 1985), which means that only 4.2 per cent of these would be served by robots. The equivalent Swedish figure was between 14 and 18 per cent.

In the case of stock in the UK, there was a total of 213 robots serving machine tools in 1984. The total stock of NCMTs amounted to approximately 32,500. Given the same assumption as above, the share of the NCMTs served by robots would be at most 1.6 per cent whilst the Swedish share is in the order of 6.7 to 8.3 per cent.

Thirdly, the Boston Consulting Group (BCG) makes the following statement as regards the demand for systems in the four major EEC countries: 'Systems [flexible cell and flexible manufacturing systems] accounted for less than 3 per cent of demand in metal cutting...' (1985, p. 3). In the case of Sweden, we estimate that in the order of 23 per cent of the value of demand for metal-cutting machine tools was in the form of FMCs and FMSs. If we include also FMMs, the figure would rise to 33 per cent. (For the methods used in this estimate, see appendix to section 5.5). The BCG estimated that the share in the four largest EEC countries would rise to about 20 per cent in the late 1980s.

Hence, all three indicators suggest that developments in Sweden are not representative for the OECD countries as a whole. The example of Sweden can, however, be used to point in the direction that other European countries will move in the years to come.

Appendix to Section 5.5:   Sources for and Methods of Estimating the Data Presented in Section 5.5

Sources for table 5.2

COLUMN (1)

NCMTs: The stock figure is based on (i) Halbert (1985a), who suggests that there were 6,010 NCMTs at the end of 1984; (ii) an assumed growth rate of the stock of NCMTs of between 11 and 12 per cent annually between 1980, when the stock was 3,900 (Elsässer and Lindvall, 1984, p. 59), and the end of 1984.

The stock figure for FMCs is based on (i) Hjelm (1985), who suggests that there were around 200 cells served by robots where each cell consists of two to four machine tools; (ii) Rizwi (1985), who suggests that ASEA has in total sold around 100 robots serving machine tools (excluding presses) in Sweden. He also suggests that ASEA has 50 per cent of the Swedish market for such robots.

The stock figures for FMS, where we set a range between 15 and 20 in the case of Sweden, are based on (i) Hjelm (1985), who suggests that there were around

20 FMSs in Sweden where the handling of the workpieces is done by automatic guided vehicles (AGVs). Through using AGVs, the flow of materials can go in any direction within the system if it consists of several cells or machines. Of course, it is quite possible that an AGV can serve only one cell with workpieces that are taken from, for example, an automatic warehouse; (ii) Steinhilper (1985), who suggests that Sweden has 15 FMSs.

COLUMN (2)

We assume that between 2 and 2.5 NCMTs are, on average, included in a FMC and that, on average, six NCMTs are included in an FMS.

COLUMNS (4) AND (5)

In order to estimate the rate of diffusion of NCMTs and systems, a survey was made of the sale of NCMTs in the period September 1984 to August 1985. We contacted the distributors and producers of 19 different brands of machining centres and ten different brands of CNC lathes plus ASEA (which dominates the Swedish robot market). We expect that these producers and distributors account for the vast majority of sales of these NCMTs. We asked the firms: (a) how many CNC lathes and machining centres did you sell during the past 12 months? and (b) how many were connected to computer controlled automatic material handling units? We included as automatic material handling units: gantry robots and dedicated robots for CNC lathes; automatic pallet changers with more than two pallets, AGVs or cranes for machining centres; and multipurpose robots. We excluded bar feeding mechanisms and gantry robots provided by the users.

In all, 246 lathes and 171 machining centres (417 units) were sold by the firms interviewed in this time period. In the UK and the USA, around 60 per cent of the flow of NCMTs is accounted for by these two types of NCMT. Assuming this to be the case in Sweden also, around 700 *NCMTs* would have been sold in the time period specified above.

In terms of *FMMs*, a total of 56 CNC lathes were sold together with some kind of dedicated material handling robot. Twenty-four machining centres with automatic pallet changer and more than two pallets were sold. Together, this amounts to 80 FMMs. Of course, there could be other types of NCMTs, e.g. CNC grinding machines, which are sold as part of an FMM, so the figure of 80 is probably an under-estimate.

On the bases of information from Rizwi (1985) we estimate that approximately 50 stand-alone robots were sold to be part of *FMCs*. We also assume that 2–2.5 NCMTs are, on average, included in an FMC.

Seventeen of the machining centres sold were to be connected to AGVs or cranes and are therefore expected to be part of FMSs. Assuming six NCMTs per FMS, we would have three FMSs. This figure is probably an under-estimation of the flow of FMSs since it excludes not only metal-forming FMSs but also those FMSs which are built by the users. One example of such is found in section 7.2.2, where the experience of Volvo Components is described.

Estimation of share of systems in total demand for
metal-cutting machine tools, in value

Approximately 130 NCMTs were sold in Sweden in the period September 1984 to August 1985 to form part of FMSs or FMCs. This amounted to approximately 19 per cent of the flow of NCMTs. Let us assume that the average price of a system is double that of the incorporated NCMTs. We further assume that the average unit price of those NCMTs that are fitted into a system is equal to those that are not. This means that in terms of *value* we need to (a) give a double weight to those 19 per cent which are incorporated into systems (2 × 19) and (b) increase the total investment by 19 per cent also (100 + 19). In such a case, the share, in value, of these systems in total demand for NCMTs would amount to 32 per cent or (2 × 19)/(100 + 19). We then need to calculate the share of systems in the total value of demand for metal-cutting machine tools. The share of NCMTs in the demand for metal-cutting machine tools in this period was in the order of 72 per cent (see n. 1 in chapter 3). Hence, the share of system demand to total metal-cutting demand amounts to 23 per cent (0.32 × 0.72). If the same exercise is repeated, including FMMs, the result is 33 per cent.

Notes

1 There is no consensus regarding these terms. For a list of definitions see ECE, 1986, p. 17.
2 Some of these FMSs may in fact be flexible transfer lines.
3 A gantry robot is a specially designed material handling robot.
4 Interviews with two leading CNC lathe producers in Japan.
5 Around 50 robots serving machine tools were sold in this time period. We assume that most of these robots are incorporated into systems that contain at least one CNC lathe.
6 These data cannot be separated from those for FMMs and would include at least some of the FMMs.
7 ASEA had 13 per cent of the world market for robots in 1981 (Bessant, 1983, p. 35). If other robot producers have, on average, the same proportion of their output allocated to serving machine tools, approximately 1,850 robots would be linked to machine tools in the world in 1981–2. The output of NCMTs in the main OECD countries in 1982 was over 40,000 units.
8 It has not been possible to distinguish between FMSs and flexible transfer lines in the empirical work presented below.
9 For an excellent description of one case see Wiehe, 1983.

Statistical Appendix to Chapter 5

Appendix table 5.1 Final products incorporating parts
manufactured by FMS

Final product	No. FMS	%
Automobiles and trucks	27	21
Machine tools	22	17
Tractors and construction machinery	18	14
Aerospace	9	7
Diesel engines	6	5
Electric motors	6	5
Pumps, valves and compressors	6	5
Hand tools, electric tools etc.	5	4
Railway equipment	4	3
Office machinery	4	3
Optical instruments	3	2
Ship engines	2	2
Material handling equipment	2	2
Sewing machines	1	
Moulds	1	
Confectionery machines	1	
Foodprocessing machines	1	
Printing machines	1	
Turbines	1	
Elevators	1	10 altogether
Air conditioners	1	
Forklift trucks	1	
Electric switchboxes	1	
Refrigeration equipment	1	
Weaving machines	1	
Coal mining machinery	1	
	127	100

Source: As table 5.1.

Appendix table 5.2 Parts/components produced by 103 FMSs

Part/component	No.	%
Housing, e.g. gearbox housings, transmission cases	46	45
Shafts, cranks and axles	13	13
Gearparts, e.g. wheels	7	7
Cylinderheads	7	7
Columns, heads and beds for machine tools	7	7
Frames, e.g. engine frames	7	7
Engine blocks	3	3
Other cylinder parts	2	2
Brakes	2	2
Compressor parts	2	2
Connecting rods	2	2
Transmissions	1	
Hydraulic components	1	
Liftarm	1	3 altogether
Hubs	1	
Valves	1	
	103	100

Source: As table 5.1.

6

Computer Aided Design Systems

6.1 The Technique and its Diffusion

Although computer support has long been available for design work, it was not until 1968–9 that an internally integrated system such as computer aided design (CAD) was marketed and put to use. The core of a CAD system is a graphic workstation at which the designer communicates with a computer in an interactive process where computer graphics plays a central role. Based on design information which is prepared in the form of geometric model products, CAD can serve as an electronic drafting board which interactively processes these geometric models. The designer is able to call forth, recreate, diversify, rotate and/or eliminate geometric elements, details or designs as presented on the terminal display screen or as stored in the computer. In this manner it is easy to recall and alter previously stored designs (Lindeberg, 1983, ch. 3).

According to one source (Lindeberg, 1983, p. 59), there were around 10,000 CAD installations in the world by mid-1982. Fifty per cent of these were located in the USA and about 35 per cent in Europe. In Sweden there were 205 systems in use (p. 64). The number of workstations connected to each system varied from one to ten or more. If one assumes that on average there were four workstations per system with on average three people using each workstation, about 120,000 designers, architects and draftsmen were using CAD in 1982, at least for some task in their work. As a comparison, there were 64,000 draughtsmen in the UK engineering industry alone in 1978 (Arnold and Senker, 1982, ch. 9.2). In the case of Sweden, it was estimated that roughly 2,500 people were using CAD in their work (Lindeberg, 1983, p. 67). The total number of designers, architects and draughtsmen was roughly 32,000 (1983, appendix 2).

When Calma, Applicon and Computervision started to market their CAD systems in 1968–9, they were based on large mainframe computers. Turn-key systems were marketed, including either their own developed computers (e.g. Computervision) or standard computers (e.g. Calma). Later mini-computers became common and in the early 1980s microcomputers or personal computers (PCs) started to be used in CAD systems. The approximate minimum cost of these computers is US$1 million for mainframe computers, US$100,000 for minicomputers and US$5,000–10,000 for PCs. The number of workstations varies from one for PC-based systems to 50 for the largest systems based on mainframe computers. Minicomputers have dominated the market over the past decade but their market share is declining to the benefit of PC-based systems as well as systems based upon large computers. Hence there is a polarization process under way (see table 6.1 which shows *hardware* costs). If, however, the number of workstations were used as the indicator instead of costs, the growth of the PC-based systems would, of course, be shown to be much higher. The development at the PC side is particularly interesting from the point of view of small and medium-sized companies. At the same time the increasing use of computer aided design (CAD) in some large engineering firms requires increasing data storage capacity for which the large computers are indispensable.

Table 6.1 Global CAD hardware cost by type of computer

	1982	1983	1986
Large computers	5%	6%	14%
Minicomputers	90%	82%	69%
Personal computers	5%	12%	17%
Total sales (US$m)	1,530	1,900	6,940

The figures for 1983 and 1986 are a prognosis.
Source: Investigation made by Jay W. Cooper at F. Eberstadt and Co., published in *The Newsletter*, no. 5, 1983 and quoted in Hansson et al., 1985.

By far the most remarkable development in CAD during the 1980s has been the appearance and growth of the PC-based systems. The process is similar to that at work in the case of CNC lathes and machining centres, which began about five years earlier. PC-based software began to be marketed in the very early 1980s. In addition to a PC and software, a plotter is needed for a complete system and the price of these components together is US$15,000–20,000. This should be compared to the high-end systems which sell for US$50,000 to 120,000 per seat (*Design Graphics World*, 1985, p. 28;[1] Moreau, 1985).

Users are finding that 'personal computer-based systems often give them 70 per cent of the benefits for 20 per cent of the cost...' (*Design Graphics World*, 1985, p. 28). This is reason enough for an explosive acceptance by the market. As a matter of fact, a whole new market segment has emerged. With an estimated 42,000 units installed globally at the end of March 1985, it is believed that already more than one-third of all CAD seats are based on PCs. The move to personal computers is expected to accelerate quickly now that more powerful machines with better displays are in the process of introduction. By 1990 'personal computer power will have increased so much that more than nine of every 10 CAD/CAM, CAE seats will use a personal or desktop computer' (*Design Graphics World*, 1985, p. 28).[2]

Now more than 25 companies globally sell PC-based CAD/CAM and CAE software for a number of applications such as mechanical and architectural draughting, finite element analysis, solid modelling and circuit simulation. Some of these companies are new in the CAD field and sell only PC-based software. Most major CAD suppliers and computer companies entered or planned to enter the market for PC-based systems during 1985. However, established suppliers of turn-key systems experience problems because users normally prefer to buy PC hardware independently or use computers already installed for other purposes. It has been estimated that users spent about US$360 million globally for PC-based CAD/CAM, CAE systems during 1984. Most of this was IBM hardware and, hence, this company seems to be a main beneficiary of the strong trend to use PC-based CAD software (*Design Graphics World*, 1985, pp. 28–9).

The leading supplier of *software* for PC-based CAD in terms of number of units sold is Autodesk Inc. This company is based in California and entered the US market in 1981. Whilst up to May 1985 they had in total sold 20,000 units of their AutoCAD system, during 1985 they were selling more than 2,000 copies (globally) of the product every month. Autodesk dominates the market, in terms of units, with a 44 per cent share. Second is Chessel-Robcom Corp., with 15 per cent (*Design Graphics World*, 1985, p. 28).

In Sweden, the marketing of AutoCAD was initiated in 1983, when Autodesk established a wholly owned subsidiary. By September 1985 about 750 systems had been sold in Sweden and about the same number in Denmark and Norway taken together (Moreau, 1985).[3]

Let us pause to reflect on these figures. At the beginning of this section we mentioned that 205 CAD systems were installed in Sweden in 1982. Again assuming an average of four workstations per system, this means 820 seats. This was before the PC-based systems were marketed. During 1983 to September 1985 one supplier alone had sold 750 PC-based systems in Sweden, i.e. almost as many workstations as the total stock in 1982. Hence the emergence of the PC-based systems provoked a virtual explosion in the diffusion of CAD in Sweden.

AutoCAD is a drafting and design system mainly for mechanical and architectural applications that runs on 30 different models of personal computer. There are, of course, important differences between PC-based CAD systems and the earlier, more sophisticated and much more expensive ones. The latter can handle much larger volumes of information, carry out solid modelling, centre of gravity calculations etc. In addition, the large computers are much faster, e.g. a reply time of one second for these as against 60 seconds for the smaller ones (Moreau, 1985).

Hence AutoCAD is currently more or less a drafting implement; very cheap at the low performance end. However, the system is gradually being improved to include other functions such as the handling of non-graphic information like component vendors and their prices (Moreau, 1985). Hence the degree of sophistication of this system is gradually increasing whilst the old CAD suppliers are designing less complex systems simultaneously. However, the latter are not as cheap as the former.

The software cost is about US$4,000 and a PC with graphics costs about US$6,000. A device for making paper copies (plotter, matrix printer or laser printer) costing about US$5,000 is also needed.[4] Hence the total minimum cost of a complete AutoCAD system is about US$15,000. With such a low initial cost the risk connected with buying and trying the new technique is negligible for most firms.[5] However, although the cost of the basic system is very low, the introduction of CAD often leads to a different mode of work in the company and thereby to related investments which may amount to several times the cost of the first system (Moreau, 1985).

6.2 The Industrial Distribution of Computer Aided Design

Of the 10,000 CAD systems installed in the world in 1982 – i.e., before the rapid diffusion of PC-based systems – between 15 and 30 per cent were outside of the engineering industry, e.g. in architectural work and in the construction and mapping industries (Lindeberg, 1983; Hatvany et al., 1981).

Within the engineering industry, data for Sweden in 1982 set out in table 6.2 show that the electrical machinery sector branch (ISIC 383) accounted for nearly 50 per cent of the systems installed. Transport equipment (ISIC 384) accounted for 27 per cent, whilst non-electrical machinery (ISIC 382) accounted for 23 per cent. The metal products share was as low as 1 per cent, probably because some of the systems used for designing building structures that could belong to the sector for metal structures were excluded from the engineering industry. For the UK, data are available on the sector-wise distribution of establishments using CAD in 1981. Broadly, the same patterns as for Sweden can be seen, with the elecrical machinery sector (383) being the largest user (see table 6.2). (It must be kept in mind that these figures relate to the period before the turbulence created by the emergence of PC-based CAD.)

Table 6.2 Establishments using CAD in the engineering industry[a]
in the UK (1981) and Sweden (1982)

Country	Metal products (ISIC 381) (No.)	(%)	Non-electrical machinery (ISIC 382) (No.)	(%)	Electrical machinery (ISIC 383) (No.)	(%)	Transport equipment (ISIC 384) (No.)	(%)	Total engineering industry (No.)	(%)
UK[b]	12	8	29	21	71	50	29	21	141	100
Sweden[c]	1	1	26	23	55	49	30	27	112	100

[a] Excluding the professional and scientific equipment sector (ISIC 385).
[b] The unit for the UK data is the number of establishments using CAD. The UK industrial classification is different from that normally used by ISIC. ISIC 381 is assumed to correspond to MHL 341, 390, 395. ISIC 382 is assumed to correspond to MHL 331–349 less MHL 341. ISIC 383 is assumed to correspond to MHL 361–369. ISIC 384 is assumed to correspond to MHL 370–384. The source suggests that there is an under-estimation of the importance of printed circuit boards manufacturers in their data, thus under-estimating the share of ISIC 383.
[c] The unit for the Swedish data is number of systems installed.
Sources: For UK, elaboration on Arnold and Senker, 1982. For Sweden, elaboration on Lindeberg, 1983.

The UK data are also available at a more detailed level. Table 6.3 shows the ten largest user industries for CAD in 1981. Again, the dominance of the electrical machinery industry is clearly marked and in particular the electronics sector. Indeed, out of the five sectors accounting for nearly 50 per cent of the total number of establishments using CAD, four were electronics-related. Apart from these applications, aerospace, motor vehicles as well as industrial plants and steel works were relatively large users of CAD. The traditional mechanical engineering sector was, on the other hand, a minor user of CAD. This was also true for Sweden, where mechanical engineering firms, apart from automobile and aerospace, used CAD only to a very limited extent. Indeed less than 10 per cent of the establishments using CAD in 1982 could be classified as traditional mechanical engineering firms.[6] (Again these data do not reflect the emergence of PC-based CAD.)

Although PC-based CAD systems are of very recent origin, some observers argue that before long nine out of ten CAD workstations will probably be based upon them (*Design Graphics World*, 1985, p. 28). Table 6.4, which shows the distribution of PC-based systems, reveals that electronic applications provide the most important user sector, as is the case for larger systems.

The use of CAD was, in the early 1980s, restricted mainly to very large firms. Out of 25 users of CAD in the engineering industry outside of the

Table 6.3 Largest users of CAD in the UK, 1981

Industry	Establishments using CAD (no.)	Accumulated share of total no. of establishments using CAD (%)
1 Radio, radar and electronic equipment	19	13
2 Radio and electronic components	18	25
3 Electronic computers	12	33
4 Telegraph and telephone equipment	12	41
5 Aerospace equipment	11	48
6 Industrial plant and steelwork	10	55
7 Other (mechanical) machinery	10	62
8 Motor vehicle manufacturing	9	68
9 Electrical machinery	5	71
10 Mechanical handling equipment	5	74
11 Other industries	38	100

Source: Elaboration on Arnold and Senker, 1982.

Table 6.4 PC-based CAD/CAM, CAE software market segment in 1984 (per cent)

	By unit shipments	By revenues
Electronic	34.2	51.5
Mechanical	14.5	13.2
Educational	24.8	17.2
A–E–C[a]	26.5	18.1

[a] Architectural-engineering-construction.
Source: *Design Graphics World*, 1985, p. 30.

electrical machinery industry in Sweden in 1982, 24 were very large firms (Lindeberg, 1983). Among 34 CAD users interviewed in Britain by Arnold and Senker (1982), only four had less than 500 employees and only eight less than 1,000 employees. Within the electrical machinery industry in Sweden, a number of smaller firms produced printed circuit boards by the use of CAD as well as the giant firms Ericsson and ASEA. (As will be discussed below, CAD is often an indispensable tool for electronics design.) In total, 11 small firms used CAD in this industry, the same, in fact, as the number of large ones.

The phenomenal diffusion of the PC-based CAD systems during the past few years means, of course, that CAD has also become available for small firms which – except for the electronics industry – did not use CAD in the early 1980s. The low cost of these systems had opened up a whole new market segment.

During 1984 approximately 250 AutoCAD software units were sold in Sweden and a similar number in Norway and Denmark together. Half of these units went to small and medium-sized firms (Moreau, 1985). Hence a breakthrough occurred in this respect during the past few years. However, this also means that large firms bought a large number of PC systems. They can be bought without previous detailed investment calculations or tough demands to show returns. In addition, funds of this order of magnitude are normally available at low levels in the firm hierarchy (Ehnqvist, 1985). It is also claimed that most of the large firms buy the small systems in order to evaluate them and that some of the firms are now considering replacing the older large-scale systems with PC-based ones (Moreau, 1985).

In this context it is of importance that the PC-based systems are easier to learn to operate than the more sophisticated ones. After one week a designer should be able to use a PC system productively and within three months he should be able to start adding his own knowledge to the system. This latter ability is important since practically all users must adapt programs to their own particular field of design (Moreau, 1985).

6.3   The Economics of Computer Aided Design

The most common reasons for introducing CAD appear to be:

1 Improved productivity of designers and draughtsmen.
2 Shortening the lead time from order to delivery or from conception to production.
3 Performing work which is too complex for manual design and drawing.

The main motives for introducing CAD vary, however, between different industries. In the electronics industry, for the production of integrated circuits, for example, the complexity of the product makes it very difficult or impossible to use manual techniques. CAD is then an essential or highly necessary tool (Kaplinsky, 1983).

In the automobile and aerospace industries, where CAD has had a relatively long application, the main concern now lies in shortening the lead time and perfecting the product design. In these industries, the cost of using CAD is quite insignificant in relation to the total production cost. To the extent that it can improve the competitive position of the firm through, for example, reducing the lead time (Giertz, 1983), CAD has been introduced without much consideration of costs.

In the mechanical engineering industry, lead time can also be of considerable importance, but, according to Arnold and Senker (1982), it does not seem to have the same general importance as in the automobile and aerospace industries. In contrast to the electronics and other above mentioned industries, CAD applications in the mechanical engineering sector seem to have two characteristics:

1 There is a larger element of experimentation. Firms acquire CAD more to learn about the technique than for immediate productivity increases (Arnold and Senker, 1982).
2 At least some of the output of the mechanical engineering industries is from machines tailored to the demand from each customer. Often these firms produce variants of the same basic design, e.g. machines of different sizes. Examples are machines used for making paper and for making biscuits.

The property of CAD of allowing the re-use of information fed into the computer implies that the design and especially the draughting work in such firms can be greatly rationalized. One early example is from the Swedish company ASEA which made a design system for different types (sizes) of transformers in 1976. This system is said to be very profitable (Kjellberg, 1984). Designing time for certain kinds of ASEA transformers was said to have decreased from 3,000 hours to 40 hours due to CAD (Alam, 1986). Designing costs for these transformers used to be an important part of the total cost of production. With such a remarkable productivity increase thanks to CAD, designing costs are no longer a significant part of total costs. Similar kinds of applications are now becoming increasingly common (Hansson et al., 1985, p. 114).

Productivity increases in the design and draughting phases are also important for firms in the automobile, aerospace and electronic industries. According to one source (Hatvany et al., 1981, p. 26) 'typical' labour productivity ratios are 4.5 to 1 for simple design and draughting with dimensioning and 2.5 to 1 for complex design with dimensioning. Arnold and Senker (1982) calculated the mean of the claimed labour productivity increases of CAD in relation to manual draughting for 24 firms and found it to be 2.74 to 1. Analysing only the firms with some years' experience with CAD led to an increase in this ratio of 3.33 to 1. Kaplinsky (1983) suggests that when CAD is used as a draughting tool the average labour productivity ratio can be 3 to 1 with variations from no increases (in the case of new drawings) to 20 to 1 (in the case of modifications of existing drawings). It should be noted, however, that these ratios measure only the work of the designers in some functions. Hence, the effect of CAD on the total labour productivity of a design department is bound to be less.

In terms of the S-curve often used to illustrate diffusion of techniques (see chapter 2) CAD technology in general could be said to have been in the introductory phase until the early 1980s. In some applications, however, electronics in particular, it had progressed into its growth phase. For most firms using CAD the first CAD was acquired in 1976–8 or even later (e.g., Sweden in 1978–80). As Arnold and Senker (1982) have shown, it can take years before the full benefits, in terms of increased productivity, can be reaped from sophisticated CAD installations. As was argued above, the mechanical engineering sector has been particularly slow in acquiring CAD and the experimental motive for buying CAD has been more important among these firms. Furthermore, only a limited number of the designers and draughtsmen in firms employing CAD normally use the CAD system (Lindeberg, 1983, p. 84). Finally, the number of firms using CAD was until recently very limited in relation to the total population of engineering firms with design activity. The high cost of investment in hardware, as well as software, may have been a reason for this.[7]

The appearance of personal computer-based CAD systems, however, seems to be changing this situation, i.e. there is now a dynamism and turbulence in the field. In chapter 2 we mentioned that the growth phase – in the S-curve – of the diffusion of a technique is characterized by changes in the technique itself as well as in the type of adopters. The technique is standardized and simplified, its price is reduced and the product is differentiated. This is exactly what has happened to CAD technology on the supply side in the 1980s. In addition, it has become much easier to learn how to use the technology. All this has led to a substantial increase in the rate of diffusion of CAD and this now includes both small and medium-sized firms.[8]

The *integration* of CAD with the other flexible automation techniques in firms is however in a very early stage of development. For example, according to Lindeberg (1983), tape or program preparation for NCMTs directly from the data in the CAD unit is very limited in Sweden. Arnold and Senker also stress this point in the case of the UK. They state, for example, that only one of the ten visited vehicle and aerospace establishments that used CAD had made significant progress using CAD for this purpose. They conclude:

> Using our strict definition of CAD/CAM (NC preparation), we have identified little activity to date. This is not in any way to suggest that CAD/CAM will not be increasingly significant in the future, but it does suggest that the process of developing CAD/CAM applications on a wide scale may take quite a long time. (1982, p. 13.13.)

This seems still to be true at the time of writing. An important reason for the limited use of CAD for NC preparation is that it can provide only part of the information required. In addition to the geometrical information supplied by CAD, data on properties of materials etc. are needed. Thus, additional data banks need to be built up in order to achieve what is, in principle, a rather simple integration between different computers (Kjellberg, 1984).

Notes

1 This excellent article, entitled 'The Shakeout Begins', is based on a study by Datatech Inc. entitled 'CAD/CAM, CAE: Survey, Review and Buyers' Guide'.
2 CAM stands for computer aided manufacturing, which implies computer-based preparation for manufacturing, e.g. NC tapes or CNC programs. There has not, however, been much progress in the direction of CAM so far and in practice CAD/CAM is actually only CAD. There are both organizational and techno-economic reasons for this, although the integration has proceeded further in electronic application than in mechanical (Ehnqvist, 1985). CAE, computer aided engineering, means largely coordination of graphic design and calculation.
3 The global figure should be around 28,000 at the same time.
4 It is important to keep in mind that hardware components have alternative applications like calculations and word-processing.
5 When a large and expensive system is bought at least US$15,000 are normally spent only on investment calculations, and evaluations. It may also be mentioned that pirate copies of AutoCAD software are sold for less than US$10 – in Singapore in 1986.
6 The large durable consumer goods firm Electrolux, which is classified as a producer in ISIC 382, had at least ten systems (Lindeberg, 1983), but this firm is not labelled as a traditional mechanical engineering firm.
7 The software packages included in the price of CAD units often have deficiencies. Therefore application-specific software has to be developed by the user, and only the largest firms have been able to afford to do that.
8 As a parallel it may be mentioned that NCMTs entered into its growth phase in the 1975–7 period.

7

Flexible Automation and International Competitiveness

In the previous four chapters we have discussed the rapid diffusion of flexible automation in the OECD countries. In the following two chapters, we will analyse two issues which arise as a consequence of this diffusion. The effects on international competitiveness within the OECD region are discussed in this chapter, whilst chapter 8 deals with the impact of the new technology on employment.[1]

We will begin here with an analysis of the quantitative impact of the new techniques in terms of their share in the investment in fixed capital in some OECD countries. The objective is to find out to what extent these new techniques are important in overall investment activity in the engineering industry. We will then put together the evidence at our disposal on the sector and product level distribution of investment in the new techniques. We conclude that the new techniques have a rising importance in terms of their share in fixed investment in machinery and equipment in the engineering sector and that for some product groups, there appears to be a greater concentration of these techniques than for others.

Two product groups where firms appear to be particularly prone to introduce the new techniques are pumps and automobile components. The experience of two Swedish firms manufacturing these products at an advanced technological level is presented. Here we also discuss the impact of the new techniques on firms' competitiveness. We conclude that there is indeed a very substantial uptake of the new techniques in these firms. This reflects that these industries are transformed into high technology ones. However, the impact at the firm level is much muted in that the functions to which these techniques are applied (design and machining) form only a part of the firms' total costs. In addition, firms' competitive strengths are a function of a range of other factors, e.g. firm-specific design experience Nevertheless, new production techniques can, if they are applied properly, give a firm the advantage of having 'super-normal' profits. Of course, 'super-normal' profits can have a significant effect on a firm's long-term performance.

Having drawn these conclusions, we proceed to analyse to what extent there is an *uneven* diffusion of the new techniques between some OECD countries. We conclude that there is a great deal of difference even between the advanced OECD countries in their uptake of the new flexible automation techniques. On the basis of a historical analysis of the case of NCMTs, we also find that the period over which firms and countries can gain 'super-normal' profits is very long indeed. Clearly then, the early adoption of new techniques can have a significant impact on the performance of firms and countries.

7.1   The Quantitative and Sector Impact of Flexible Automation Techniques

In chapters 3 to 6 we discussed the benefits that may accrue to the investors in new flexible automation techniques. Although the benefits vary in both type and magnitude, on the whole it is clear that they improve the investor's position in the market. The question then arises as to the order of magnitude of the importance of these techniques in the overall investment activity within the engineering industry. Clearly, if a substantial impact is to be expected, these techniques would have to account for a significant share of the overall investment. Let us therefore turn to this issue.

Unfortunately, a complete analysis of the share of flexible automation techniques in fixed investment in the OECD countries is not possible on account of lack of data. However, there are data for investment in NCMTs and in robots for some countries and for some years. These partial data are interesting enough to warrant attention.

In appendix table 7.1 we summarize the data available for Japan, Sweden, the UK and the USA on the share of investment in NCMTs and robots combined as a part of the total fixed investment in machinery and equipment in the engineering industry. In the case of Japan, the share of NCMTs and robots rose (see column 8) from 4.7 per cent in 1977 to 18.8 per cent in 1984. The Swedish picture is roughly the same as the Japanese, whilst that for the UK shows a somewhat lesser importance of these technologies in the overall investment in the engineering industry. Finally, the USA appears to be far behind, although there is a risk that the data somewhat understate the importance of these techniques in the fixed investment in machinery and equipment (see note (i) to appendix table 7.1). Clearly, these partial data indicate, with the exception of the USA, a sharply rising share of NCMTs and robots in the investment in machinery and equipment in the engineering industry of the OECD countries. Again, we would like to underline the incompleteness of these data. The inclusion of CAD would probably greatly increase the share of flexible automation techniques. For example, CAD for the mechanical and electronic sector was sold to a value of US$159 million in the UK in 1985 (Aronsson, 1986, p. 11). As a comparison, sales of NCMTs in the UK amounted to US$259 million in 1984.

Having established that flexible automation techniques account for a significant and rising share of investments, we shall proceed to establish which sectors and product groups are most apt to introduce these techniques. Unfortunately, sector level data on the value of investments of the various flexible automation techniques are not available. Table 7.1, however, presents a summary picture, by sector, of the distribution of the stock of NCMTs, robots, CAD and FMSs in the engineering sector of some OECD countries. A number of observations can be made on the basis of this table. First, the metal product sector (ISIC 381) is badly represented in the table, indicating a generally small importance of flexible automation techniques. Secondly, CAD is most heavily used in the electrical machinery sector (ISIC 383). This refers to both the larger type CAD units and the PC-based CAD units (*Design Graphics World*, 1985). Thirdly, the non-electrical machinery industry (ISIC 382) and the transport equipment industry (ISIC 384) are fairly well represented by at least two techniques. The former has a large part of both NCMTs and FMSs, while the latter has a large part of both robots and FMSs.

Table 7.1 Summary of the sector-wise distribution of various flexible automation techniques (percentage of stock, in numbers)

Technology (country)[a]	381	382	383	384	385
NCMTs (USA)	15	54	11	16	5
Robots (FRG)	8	15	18	59	n.a.
CAD (UK)	9	21	50	21	n.a.
FMS (global)	5	52	5	36	2

[a] We have chosen to present the data from different countries as we do not have, apart from Sweden, a complete set of data on NCMTs, robots and CAD for one country. The NCMT pattern of diffusion is, however, much the same in other countries as in the USA; the pattern for robots in the FRG is repeated in the UK, and the pattern for CAD in the UK is more or less repeated in Sweden. We would therefore suggest that the table should provide a reasonably good approximation of the global pattern of diffusion.

Source: NCMTs, table 3.4. We have excluded those NCMTs classified as being used in primary metals and in miscellaneous production. Robots, table 4.3. CAD, table 6.2. FMS, table 5.1.

In terms of product groups affected by the diffusion of these techniques, the high share of robots and FMSs in the transport equipment sector reflects to a great extent the *automobile industry's* rationalization programmes in the 1980s. One of the most important applications of robots is still spot welding (appendix table 4.3), mainly in the production of automobiles. Assembly robots are also being diffused to this industry. An important share of the FMSs are installed in firms producing automobile components, which is evidenced, for example, by the case of Volvo diesel engine production in Skövde, Sweden, where three different FMSs are working (see section 7.2.2). Large CAD systems as well as PC-based CAD units are normally also found within this industry. Hence, in this industry, we can see a convergence of all the four technologies. Similarly, in the *heavy electrical equipment* industry there is a convergence of all four techniques.

Other products also show an impact from more than one of these techniques. Although not claiming to be exhaustive, the following list of product groups would be found among those most affected by the application of flexible automation techniques. *Aircraft and parts* is traditionally an industry which is at the forefront of application of new technology. This applies also to flexible automation techniques, apart from robots. *Cutting tools, pumps, valves and compressors* are other products that use several flexible automation techniques. In the case of cutting tools we are mainly talking about NCMTs, robots and CAD and in the latter case the techniques in question also include various types of FMSs. For a case study of pumps see section 7.2.1.[2] *Construction machinery, tractors and machine tools* have a high share of the stock of both NCMTs and FMS. Some firms are also beginning to use CAD. *Special industrial machinery* is also a large user of NCMTs and this sector is also beginning to use CAD. *Electronic components* are large users of CAD and have some special NCMTs. For the products listed above the impact of flexible automation techniques, in terms of their share of fixed investment in machinery and equipment, would be expected to be higher than is indicated by appendix table 7.1, which reveals the impact for the entire engineering industry.

7.2 The Impact at Firm Level –
Evidence from Pump and Diesel Engine Manufacturing

Two product groups where firms have taken up the use of flexible automation in an extensive way are pumps and automobile components. In this section we will discuss the firm level impact in two Swedish firms, both world leaders in their respective areas. The firms are Flygt AB, in the field of submersible pumps, and Volvo Components which produces, amongst other things, diesel engines.

7.2.1 FLYGT AB

A submersible pump (SP) is one where the motor itself is placed in the liquid that is to be handled by the pump. Flygt AB was one of the first firms to produce SPs and is the world leader in the field. Flygt has more than 50 per cent of the world market for SPs and its market share is even higher for particular types of SP. Flygt has around 3,000 employees and a sales value of some US$200 million. It is a full line supplier of SPs whereas many of the other firms in the industry cover only a part of the spectrum of SPs. Due to its ambitions of continuing to be a full range supplier in the industry, Flygt has developed a very large design staff. Indeed, its design staff of about 100 persons is four times that of one of its main, Japanese, competitors (Tsurumi). This is all the more necessary since Flygt is pioneering extended applications for the technology of SPs. Indeed, 45 per cent of their products are less than three years old. Keeping the leadership within those segments of the industry that are more conventional, e.g. sewage pumps, and are characterized by little or no patent protection and relatively low barriers to entry, means that Flygt has to put a lot of emphasis on lead time reduction, cost efficiency and marketing.

Most of the 70,000 SPs that Flygt produces annually are made in one factory in Sweden. This factor has around 1,000 employees and the remaining approximately 2,000 employees are involved in marketing and after-sales service, mainly abroad. More than 90 per cent of the production is exported. Roughly one-third of the sales value of the firm is generated in the factory making the SPs; another third is generated by Flygt's subsidiaries and the remaining third is bought items and services needed by the subsidiaries in their sales activity.

The new techniques that we are studying are located mainly within the production unit which accounts for one-third of the sales value. Within this factory, value added accounts for 60 per cent of production value. Hence, if the sales value in total amounts to US$200 million, approximately 40 million is value added in the production unit. It is primarily these 40 million that new manufacturing techniques can reduce and we can immediately see that even a major increase in productivity in the value added process will be much diluted if other components making up the total costs stay constant. However, due to increasing competitive pressure arising from new entrants in the industry, Flygt is heavily emphasizing the application of new technology to stay cost-competitive. Although such efforts result in a marginal reduction of costs in relation to its total sales value, it may nevertheless be of great significance in relation to profits. Indeed, Flygt is the most profitable firm in the industry and its continued efforts to reduce production costs are very probably one significant cause of its excellent profit performance. We will briefly describe below the efforts of Flygt in improving cost efficiency through using flexible automation techniques.

Flygt has two (large) Computervision CAD units with nine workstations in total. One has CAM functions, i.e. it is partly used to translate designs to programs for the NCMTs. The firm also has a number of PC-based CAD units. The use of CAD entails a number of advantages, of which perhaps the most important is reduction in the lead time from conception to production. This applies to both custom designed products, where this advantage is more obvious, and to standard products.

Let us elaborate on the importance of CAD in reducing the lead time in the design of standard SPs. Standard pumps are produced in a range of varieties, e.g. in terms of sizes. The ability of CAD to reduce the lead time means that Flygt can make *all* design variants and sizes practically speaking simultaneously and thus market a whole new series of pumps simultaneously. A competitor without CAD then has difficulties in copying all the variants quickly, due to the sheer amount of drafting required. Other benefits, such as increased productivity of the designers, are also of importance.

Flygt began investing in NCMTs in 1968 and now has 36 units. Some 50–60 per cent of the firm's annual investments are normally made in the form of NCMTs, which is very high compared to the average for the engineering industry (see appendix table 7.1). From 1977 Flygt bought no conventional machine tools and these are no longer even considered as an alternative! Of the NCMTs, five are machining centres and two of these are parts of flexible manufacturing modules (FMMs) in that they are equipped with automatic pallet changers and have more than two pallets. In addition, Flygt have a very large combined turning, milling and boring machine which is also part of an FMM.

A number of the NCMTs and FMMs are located in a new workshop which is specially designed to produce smaller and medium-sized sewage pumps. The motive behind investing in the new workshop was to solve the problem of being able to meet the short delivery times demanded by the market without having excessive stocks of those products produced for anticipated demand. The workshop is organized along the principles of group technology which means that all the machine tools needed to produce a particular part are placed together. Therefore a component is quickly run through all the stages of machining. Lead time can therefore be reduced from weeks to days. Other advantages with this organizational principle are simplified planning and that one operator can handle several machine tools.

In terms of the technique used, no conventional machine tools exist in the plant. One advantage with using NCMTs is that the set-up time is virtually zero when a part that has been produced earlier, and thus for which there is a computer program already made, is produced again after a time interval. This further reduces the need for carrying stocks. In addition, the information links between machining, stocks, assembly and testing are all computer-based. It is paper-less production which means that there has been a considerable saving in white collar workers. Thus, we have here a flexible and highly automated factory for the machining of these pumps, which are produced in quantities of 35,000–40,000 units annually. Through the implementation of the principle of group technology in combination with advanced techniques, Flygt can reduce costs of production (through low labour costs, lower costs of fixed capital in the cases of at least some systems, and lower costs for stocks and work in progress) and meet the demands for short delivery times.

The workshop is connected to a superior computer which in turn has on-line links with Flygt's 17 subsidiaries abroad. One aspect of this integration is that they have been able to reduce their global stocks – i.e. including stocks in subsidiaries – of spare parts by approximately 20 per cent. They are now planning the extension of this system to complete pumps. Note that the savings here also affect the costs outside of the production unit in Sweden.

In all areas of the design, production and marketing fields, there is therefore a considerable introduction of new technology and, furthermore, the various 'islands of automation' have to an important extent been joined through a central computer. Flygt is in the process of approaching CIM, i.e. computer- integrated manufacturing, which we referred to in chapter 1.

7.2.2 VOLVO COMPONENTS

Volvo Components (VC) is a wholly owned subsidiary of the Swedish automobile and truck producer Volvo. VC has several factories producing components and three of them produce diesel engines. The diesel engine division has 1,500 employees and supplies diesel engines to trucks, buses, construction machinery and boats. For large diesel engines, that is for trucks over 16 tons, VC is the third largest producer in the world.

One of the three diesel engine factories is at Skövde, where 60,000 diesel engines of between 6 and 12 litres capacity are manufactured annually. VC produces a number of components in-house, such as cylinderheads, engine blocks, crankshafts, exhaust pipes etc. In addition, VC assembles the engines. The value added amounts to 40 per cent of the production value. Around 15 per cent of the production value is accounted for by the assembly process which leaves 25 per cent for machining, overheads etc. These are the areas mainly affected by flexible automation techniques. In order to reduce unit production costs, VC has put a lot of emphasis on the application of flexible automation techniques in the past few years. We give an account of some of these efforts below. CAD is used to only a very small extent by

VC since the mother company provides them with all the designs. (The mother company makes extensive use of CAD.) VC in Skövde has, however, one PC-based CAD station for the purpose of designing tools and fixtures.

As far as the machining technology is concerned, some components, e.g. cylinderheads, are produced in fixed transfer lines where the setting time is 20 hours. Other articles are produced using flexible manufacturing systems of various kinds. In future, VC will depart further from using fixed transfer lines in favour of flexible automation.

Since 1979, there has been a virtual 'technology explosion' at VC. Today the company does not buy any manual machines and in the past ten years it has not bought any transfer line. In the section where they produce axles, shafts, general articles etc. they have recently introduced flexible techniques to a large extent. They have here around 50 NCMTs and ten industrial robots. In addition, they have a large number of gantry robots which VC have built themselves. Two of the industrial robots are process robots and eight serve machine tools with the handling of workpieces. All robots form parts of FMCs or FMSs. All in all they have three FMSs and seven to eight FMCs. They also have a few FMMs which consist of machining centres with pallet pools. We present some examples of this new technology below.

Example 1: Production of gear case tops   Some components for the diesel engines are traditionally looked upon as very difficult to produce. Conventional wisdom has regarded special machine tools and expensive fixtures as necessary for this type of production. The old VC technology for gear case tops consisted of 16 special and standard machine tools connected to each other by means of conveyors. The loading and unloading of the workpieces was done manually and each workpiece had to pass through all the machines. The new technology consists of three machining centres, two other machines, two robots and a number of automatically guided vehicles (AGVs). This system is connected to three other FMCs through the means of the AGVs. The investment cost of the system (not including the additional three FMCs) was in the order of US$2.5 million. In the pre-investment calculation, VC compared the new system with buying new machine tools of the dated type in addition to renovating those that are no longer available in the market. The increased capital cost (around 30 per cent) was well compensated for by reduced labour costs. The (pre-investment) payback period was estimated to be 3.2 years.

Example 2: Rough machining of large crankshafts   In this system two people per shift handle 11 machine tools, e.g. two CNC lathes and one machining centre. The machines are served by gantry robots and they also have a number of AGVs. The superior computer communicates with the AGVs and the gantry robots. The robots, in turn, communicate with the machine tools. In this system, they have reduced the number of operators by 60 per cent and also reduced somewhat the work in progress.

In the *complete* crankshaft factory, the introduction of new technology reduced the number of machines from 54 to 41. The cost of buying the old machines today would have been 76 million Sw.kr whilst it was 89 million Sw.kr for the new machines. The production capacity of the new machines, however, is 33 per cent larger than for the old machines, which means that the new technology is capital-saving as long as the utilization ratio is kept high. The machines have also reduced the labour cost substantially. For the 54 machines 90 workers were needed in total for two shifts; with the new machines 40 workers are needed in total for three shifts. Hence, the new technology is both capital- and labour-saving.

Example 3: Production of exhaust pipes This FMS consists of four machining centres, four preparation stations, a stock of raw materials and fixtures, two magnetically guided fork lift trucks and a superior computer. Three persons are employed per shift and the cost of the systems was around US$2 million. Thirty different variants of exhaust pipe for four engine families are produced. All machines can produce all types of exhaust pipe and the system is therefore a fully flexible one – or 'chaotic' as the production people at VC call it. VC has had considerable 'running' problems with this FMS, partly due to poor software delivered by a specialist firm.

Example 4: Production of small components For smaller components, VC try to follow the principle of group technology and they have established some seven to eight FMcs for the production of, for example, cog wheels.

VC has to a very considerable extent begun to implement the 'factory of the future', although it should be noted that the implementation of these systems has certainly not been without problems. Lessons can be drawn as regards the need to adjust the various skills of the employees, at both the level of factory workers and that of production planners, and also among the suppliers of the system.

7.2.3 NEW PRODUCTION TECHNOLOGY AND COMPETITIVENESS

These examples show that pump and diesel engine manufacturing are gradually being transformed into high technology industries. In the area of *design*, CAD is being extensively used; in *machining*, NCMTs, FMMs, FMCs and FMSs (including robots) are normally employed; in *internal transport*, AGVs are sometimes used; in *inventory control*, computers have long been used and in *marketing*, computers are being connected to a superior computer. Whilst the advantages of using these techniques are clear, the question remains what conclusions can we draw for competitiveness at the level of the firm and country from an early introduction of these techniques.

Unfortunately, much of the literature on the impact of new techniques on international competitiveness proceeds as if the determinant of the competitiveness is fully or mainly a question of relative production costs. New automated microelectronic controlled production techniques have, for example, been seen as a threat to the developing countries in that they have the potential to erode their competitive advantage in some traditional industries such as garments and shoemaking where some developing countries have achieved a good performance, partly on the basis of an input of cheap labour. Similarly, flexible automation has been seen as a saviour of, for example, US industry: 'High tech to the rescue – more than ever, industry is pinning its hopes on factory automation' (*Business Week*, 1985, p. 84).

This focus on production costs, and on factor prices, has its intellectual origin in the factor proportion theory of international trade. As is well known in the literature (Gray, 1980), this theory is inadequate in explaining existing trade flows, and therefore firms' and countries' international competitiveness. The Technology Gap trade theory (Hufbauer, 1966; Soete, 1981), as well as much literature on the border between economics and management (Caves, 1980),[3] emphasize, in contrast, a range of other factors. Firms are seen as striving to create unique assets which make them superior to their competitors. These superior, firm-specific assets are basically dynamic in character and are, hence, augmented by the accumulation of experience, in research and development, design, production and marketing. Static economies of scale can further reinforce and prolong any advantages derived from an early technological breakthrough or from a faster accumulation of experience in any or all of the functions in the firm.[4]

Thus, production costs, which are partly influenced by the techniques adopted, is only one out of many determinants of competitiveness. These can include:

- Low production costs due to an early adoption of new technologies or favourable factor prices.
- Firm-specific knowledge and skills in, for example, design and in bargaining with suppliers.
- Product-specific factors, such as fixed economies of scale.

In the cases discussed here, Flygt and Volvo Components, we can make several observations as regards the impact of the new techniques on competitiveness.

Firstly, the advantages to the adopters of the new technique are not limited to a reduction in production costs but are also shown in a reduction in lead times and in the ability to design better products.

Secondly, the modern firm in the engineering sector finds that a great deal of its costs refer to activities other than production proper. This is seen most clearly in the case of Flygt, where costs of production proper were only one-fifth of the sales of the company. Hence, techniques introduced to reduce the costs of production proper will have only a marginal impact on the firm's overall costs to supply the customer with the product.

Thirdly, again in the case of Flygt, the survival of the firm rests on a range of factors other than its ability rapidly to introduce new techniques. Flygt was a pioneering firm in the field of submersible pumps and already in the 1950s and 1960s it expanded internationally and built an extensive marketing network abroad. Through this network, it now enjoys consider- able benefits from economies of scale. For its smaller competitors the absence of such a network acts as a barrier to entry into the various markets.

Flygt also enjoys other benefits from the static scale economies that it reaps owing to its dominant position in the industry. For example, raw material costs can be reduced through bulk purchases, design costs can be spread out on a greater number of units produced etc. Furthermore, Flygt has, with its emphasis on design development, created significant firm- specific skills. This applies in particular to the knowledge involved in applying the basic SP technology to new application areas. Due to its early entry into new application areas, it is probable that important dynamic scale economies[5] will continue to arise within the firm as experience about these new areas accumulates. The Technology Gap account of trade, where an initial technological breakthrough gives advantages that are prolonged by static and dynamic economies of scale, can well be applied in this case, where Flygt's early entry into the industry still gives it considerable advantages. Hence, static economies of scale, firm-specific skills and dynamic economies of scale figure as important determinants of the international competitiveness of Flygt, in addition to Flygt's use of new techniques.

As these examples show, there is a range of factors that need to be considered in analysing the international competitiveness of firms and countries. The production technology used is certainly only one out of many determinants of competitiveness.

In chapters 3 to 6, however, we have seen that the introduction of flexible automation techniques is normally quite profitable and in certain specific applications these techniques can be highly profitable. Let us repeat some examples from earlier chapters. NCMTs may reduce the cost of production of certain parts by 50 per cent (section 3.4). The pay-off period of industrial robots has been shown to be two to three years (section 4.5). Although FMSs are still an immature technique their cost reduction may be substantial (sections 5.4 and 7.2.2). In the case of CAD the productivity of designers may increase by several hundred per cent in certain applications (section 6.3).

Therefore, although the cost efficiency of design offices and plants cannot always, or not even often, be expected to be *the* decisive factor determining the competitive strength of firms, one can on fairly safe grounds assume that the *profit level* is highly sensitive to the introduction of new techniques and to the skill with which managers and workers apply it. That is, at the margin, new techniques can have a significant impact on the competitiveness of firms and countries.

7.3 A Comparison of the Level of Diffusion of Flexible Automation Techniques in some OECD Countries

In the previous sections we have reached the following conclusions:

- The share of flexible automation techniques in investment in machinery and equipment is significant and rising in the engineering industry.
- The share is considerably greater among some product groups than in others, that is, there is an uneven diffusion of the techniques between product groups.
- At the firm level in two product groups highly affected by flexible automation techniques – pumps and automobile components (diesel engines) – the impact of the new techniques is overwhelming in that these firms do not acquire *any* of the traditional techniques any more.
- The impact of the new techniques on the overall performance of the firms is, however, less significant, since they affect only a few dimensions of the competitive strength of the firm, chiefly *parts* of the production costs and lead time.

Although the effects of the new techniques are muted in this way, the lead that an early adopter can get may be of great importance. After all, if a firm has higher than average profits for some time due to an early and successful application of new techniques, the long-term survival and development of the firm will be affected in a positive way. It then becomes of interest to analyse to what extent some individual OECD countries adopt the new techniques faster than others and to assess the length of the period during which an early adopter can gain 'super-normal' profits on account of the new techniques.

In table 7.2 we show the present density of the four techniques discussed within five OECD countries. The measure used is the number of units of each technique installed divided by millions of employees in the engineering industry. From the table, two main observations can be made. First, among these OECD countries, Japan and Sweden are far beyond the other three in terms of the density in use of these techniques. The FRG, UK and USA are broadly speaking at the same level. The exception is for CAD, where the USA is on a par with Sweden and Japan is broadly at the level of the FRG and UK. The relatively weak position of CAD in Japan is probably due to the same reason that office automation equipment in general is diffusing only slowly in Japan, namely the problems associated with using the Japanese language in computers. In general, the diffusion of office automation in Japan is far behind that of the USA (Sasaki, 1987).

Secondly, there is a considerable discrepancy between the four techniques as regards the extent of the differences in density:

- For NCMTs the density varies between 10,505 in the UK and 22,399 in Japan, i.e. by a factor of 2.1.
- For CAD it varies between 1,380 in Japan and 7,011 in Sweden, i.e. by a factor of 5.1.
- For robots it varies between 846 in the UK and 12,257 in Japan, i.e. by a factor of 14.5.
- For FMSs the density varies between three in the UK and 55 in Sweden, i.e. by a factor of 18.3.

The measure of density used is certainly imperfect and the data base is not very solid. Hence, the results can only be seen as approximations. Still, some further observations can be made.

The difference in density is clearly largest for those techniques that are least mature, i.e robots and FMSs. NCMTs and CAD systems have already proceeded quite far along their S-curves.

For the less mature techniques, i.e. robots and FMSs, Sweden and Japan have indeed a very considerably higher density than other countries. To the extent that these production techniques mean a lower cost of production and shorter lead times, (the firms using these techniques in) these countries will profit from a competitive advantage in relation to (firms not using them in) other countries. As is suggested by relatively small differences in the densities between the five countries with respect to the more mature technologies, especially NCMTs but also CAD, it is, however, probable that these differences will be evened out when robots and FMSs become mature, i.e. during the next 10–20 years. Hence, the competitive advantages based upon differences in diffusion of production techniques in OECD countries would be largely temporary in character. The period in which early adopters can gain 'super-normal' profits can be expected to vary between techniques.

The hypothesis that the competitive advantages mentioned are only temporary in character is supported by the history of diffusion of NCMTs. The case of NCMTs may also suggest to us the length of time that it may

Table 7.2 The stock and density[a] of flexible automation techniques in a number of OECD countries

	(1) NCMTs	(2) Robots	(3) CAD	(4) FMS
FRG				
stock	46,435 (1984)[b]	6,600 (1984)	11,000 (1983)	25 (1984)
density	11,376 (1980)	1,617 (1980)	2,694 (1980)	6 (1980)
Japan				
stock	118,157 (1984)[b]	64,657 (1984)est.	7,300 (1984)[c]	100 (1984)
density	22,399 (1980)	12,257 (1980)	1,384 (1980)	19 (1980)
Sweden				
stock	6,010 (1984)	1,900 (1984)est.	1,900 (1984)[d]	15 (1984)
density	22,177 (1980)	7,011 (1980)	7,011 (1980)	55 (1980)
UK				
stock	32,566 (1984)	2,623 (1984)	9,000 (1983)	10 (1984)
density	10,505 (1980)	846 (1980)	2,903 (1980)	3 (1980)
USA				
stock	103,308 (1983)	13,000 (1984)est.	59,400 (1984)[e]	60 (1984)
density	11,728 (1980)	1,475 (1980)	6,743 (1984)	7 (1980)

Table 7.2 notes

a Number of techniques divided by million employees in the engineering industry.

b Nagao (1985) gives stock figures of 32,000 for the FRG in 1982 and 56,400 for Japan in 1981. On the basis of data from VDMA (1985), and National Machine Tool Builders' Association (various), we have updated these figures. We have simply added the apparent consumption of NCMTs for the subsequent years to Nagao's figures. The Japanese figures for the years 1982–4 exclude metal-forming NCMTs.

c The figure is estimated by the authors on the basis on the *Japan Economic Journal*, 1986, p. 147, which shows the annual sales of various types of CAD units in Japan. We have assumed that mainframe CAD units are on average equipped with four workstations and that there were 700 mainframe CADs in Japan in 1984. We further assume that all other CAD units have one workstation each.

d The stock in 1982 was 208 units (Lindeberg, 1983). We have assumed four workstations per system and an annual growth rate of 50 per cent of work-stations. This is probably a conservative estimate since AutoCAD alone sold 750 PC-based CAD units in the period 1983 to September 1985.

e Arnold (1984) estimates that there were 6,600 systems installed in the USA in 1982. If we assume an average of four workstations per system and a 50 per cent growth rate per annum, the stock in 1984 would be 59,400. This estimate appears to be reasonably accurate. Aronsson's report (1986) of a study in the USA made in 1985 claimed that there were 15,000 CAD systems in the USA manufacturing industry. If, on average, there are four workstations per system, the number of workstations would be 60,000. On the other hand, the figure might not include PC-based units which would result in an under-estimate of the stock of CAD.

Sources: NCMTs: for Germany and Japan, Nagoa, 1985 and note (b); for Sweden, Halbert, 1985; for the UK and USA appendix table 9.2. Robots: table 4.2. CAD: for the FRG and UK, Northcott et al. 1985; for USA, note (e) above; for Sweden, see note (d) above; for Japan, see note 10. FMS, Steinhilper, 1985. The number of employees in the engineering sector is taken from table 4.2.

take for late adopters in the OECD countries to reach the density figures of the early adopters. The figures on stocks of NCMTs as well as densities in the use of NCMTs for the FRG, Sweden, the UK and USA are presented in table 7.3 for selected years between 1967 and 1984.

Table 7.3   The stock and density[a] of NCMTs in a number of OECD countries

	1967	1970	1976	1984
FRG				
stock	730	1,930	6,644[b]	46,435
density	177	467	1,608	11,376
Sweden				
stock	220	400	2,100	6,010
density	797	1,449	7,609	22,177
UK				
stock	1,500	3,200	9,725	32,566
density	417	890	2,704	10,505
USA				
stock	12,000	20,000	40,492[c]	103,308 (1983)
density	1,645	2,741	5,551	11,728

[a] The density has been calculated as the stock of NCMTs, in units, divided by millions of employees in the engineering industry in 1971 for 1967, 1970 and 1976. The latter figure was for the respective countries: FRG 4,132 (1972); Sweden, 0,276; UK, 3,596; USA, 7,295. The source for these figures was ILO, 1981, table 5B. The density figures for 1984 were, however, calculated using the 1980 data on employment, see table 4.1.

[b] This figure is estimated. We assumed that the average annual cumulative growth rate of the stock of NCMTs of 22.9 per cent between 1970 and 1974 (when the stock was 4,400 according to Steen, 1976) applied also to the period 1974–6.

[c] This figure is estimated. We assumed that the average annual cumulative growth rate of the stock of NCMTs was 11.3 per cent between 1973 and 1976. This was the rate in the period 1973–8, as can be seen in appendix table 9.2.

Sources: Nabseth and Ray, 1974, p. 31 for stock of NCMTs in 1967 and 1970. Table 7.2 for stock of NCMTs in 1984. Steen, 1976 for stock of NCMTs in Sweden and USA in 1976. For stock of NCMTs in the FRG in 1976, see n. (b). Appendix table 9.2 for stock of NCMTs in the UK in 1976.

In 1967 the density varied between 177 in the Federal Republic of Germany and 1,645 in the USA, i.e. by a factor of 9.3. In 1970 the density varied between 467 in the FRG and 2,741 in the USA, i.e. by a factor of 5.9. In 1976 the difference in densities between Sweden, which then had surpassed the USA, and the FRG amounted to a factor of 4.7. Finally, in 1984, the difference between Sweden and the UK, which now has the lowest density, came to a factor of 2.1. The initial, very large difference in the density of NCMTs gradually gave way over this 17-year period to a situation where the national differences were not as substantial as before. The difference is, however, still almost 2 : 1, which is surprisingly high.

It is also interesting to note that the density was highest in the USA in the early period, i.e. it was highest in the country where NCMTs were first innovated. However, the USA lost its leading position in terms of density to Sweden some 20 years after the time when the first NCMT was produced (in the early 1950s) and a decade after they began to be diffused on a noticeable scale (in the mid-1960s). This points to the importance of national suppliers in the initial stage of the diffusion of a new technique, as argued in the discussion of the S-curve in chapter 2.[6] The data also suggest that it may take towards 20 years before a new technique is relatively evenly diffused among countries that are so similar as these OECD countries. This is also the time period under which the early adopters can gain 'super-normal' profits from the early introduction of the new technique. Thus, although relative production costs constitute only one of several determinants of the competitive strength of firms and countries, the period under which 'super-normal' profits can be gained by early adopters is very long. An early successful application of a new technique can therefore be of some importance to the long-run performance of firms and countries.

On the basis of the NCMT case, it can also be noted that with regard to differences between countries in terms of FMS and robot densities factors of 18.3 and 14.5 are not excessively high, although higher than the case of NCMTs in 1967. However, we would emphasize that to the extent that the emergence of FMSs marks the beginning of a broad and deep process of change, the countries and firms that take a lead in this field may well profit from an advantage based thereon for an even more extended period of time than was the case of NCMTs. One reason is that FMS is not, like NCMTs, a question of only machining in stand-alone units, but a much more 'comprehensive' and also more complex process of change. It is a question of building extended production *systems* which may also be connected to CAD, inventory control, marketing etc. The complexity of this task means that the process of diffusion of FMSs is to be expected to be fraught with more difficulties than that of stand-alone NCMTs. In addition, since the economic benefits of FMSs apply to large sections of a plant, and in the future maybe a whole plant, the 'super-normal' profits of early adopters may well be relatively greater than those connected with an early adoption of NCMTs.

Notes

1 The effects of the new technology on the competitiveness of the OECD countries as a group in relation to the developing countries will be discussed in chapter 12.
2 Bessant and Hayward (1986) report four FMSs or FMCs used for pump production in the UK.

3 As Caves puts it in a review article on the concepts and theoretical framework of the positive economics of corporate strategy: 'The firm rests on contractual relations that unite and coordinate various fixed assets or factors, some of them physical, other consisting of human skills, knowledge, and experience – some of them shared collectively by the managerial hierarchy. These factors are assumed to be semipermanently tied to the firm by recontracting costs and, perhaps, market imperfections. An...implication of these heterogeneous fixed assets is that the firm can succeed...in a given market by possessing superior assets of any of several types. Equally successful market rivals thus may employ quite different bundles of fixed asset qualities...The standard model of perfect competition assumes these fixed factors away' (1980, p. 65).

4 In Hufbauer's words: 'Technology gap trade is therefore the impermanent commerce which initially arises from the exporting nation's industrial break-through, and which is prolonged by static and dynamic scale economies flowing from that breakthrough' (1966, p. 29). As Jacobsson (1986, ch. 3) shows in the case of computer controlled machine tools, the technology gap trade does not necessarily have to apply to the introductory phase only. Indeed, in this case it was the Japanese producers who, on the basis of new design concepts introduced in the *growth* phase of the product life cycle, could capture benefits of economies of scale ahead of their competitors and on this basis create a long-term competitive edge.

5 Dynamic economies of scale refer to the firm-specific benefits that come out of an accumulation of experience in the firm, be it in the design, production or marketing spheres. Static economies of scale are product-specific and result in a reduction in the unit cost as production volume per unit time increases, at least up to a certain point.

6 The relatively strong position of the US in terms of CAD diffusion is probably connected to the US dominance in the CAD supplying industry.

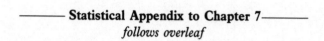
————— **Statistical Appendix to Chapter 7**—————
follows overleaf

Appendix table 7.1 Estimated share of investments in NCMTs and robots in the total investments in machinery and equipment in the engineering industries of Japan, Sweden, the UK and the USA, 1977–84

(1) Year	(2) Total fixed investment in machinery and equipment in the engineering industry (US$m)	(3) Investment in robots (US$m)	(4) (3)/(2) (%)	(5) Investment[a] in NCMTs (US$m)	(6) (5)/(2) (%)	(7) (4)+(6) (%)	(8) True share of robots and NCMTs in investment in equipment[b] (%)
Japan							
1977	6,291	35	0.6	169	2.7	3.3	4.7
1978	8,182	64	0.8	227	2.8	3.6	5.2
1979	9,990	93	0.9	495	5.0	5.9	8.2
1980	14,693	225	1.5	733	5.0	6.5	9.6
1981	17,095	356	2.1	973	5.7	7.8	11.5
1982	15,075	507	3.4	1,066	7.1	10.5	15.6
1983	15,967[c]	671	4.2	1,166	7.3	11.5	17.4
1984	18,913[d]	806	4.3	1,567	8.3	12.6	18.8
Sweden							
1977	616	8	1.3	31	5.0	6.3	9.3
1978	600	8	1.3	34	5.7	7.0	10.2
1979	757	14	1.8	48	6.3	8.1	12.0
1980	1,008	n.a.	n.a.	65	6.4	n.a.	12.6[e]
1981	942	n.a.	n.a.	64	6.8	n.a.	16.3[e]
1982	750	10	1.3	56	7.5	8.8	19.7[e]
1983	703	9	1.3	95	13.5	14.8	16.7[f]
1984	867	15	1.7	120	13.8	15.5	17.8[f]

UK

Year							
1978	3,096[g]	n.a.	n.a.	148	4.8	n.a.	6.9[h]
1979	4,320	n.a.	n.a.	247	5.7	n.a.	8.0[h]
1980	4,549	10	0.2	344	7.6	7.8	11.0
1981	3,263	20	0.6	259	7.9	8.5	11.1
1982	3,010	26	0.9	221	7.3	8.2	11.7
1983	2,879	27	0.9	205	7.1	8.0	11.6
1984	3,092	39	1.3	259	8.4	9.7	13.9

USA

Year							
1982	30,163[i]	185	0.6	1,557	5.2	5.8	8.3
1983	29,544[i]	79	0.3	973	3.3	3.6	5.1
1984	37,474[i]	225	0.6	1,286	3.4	4.0	5.8

[a] For Japan, investment in metal-forming NCMTs is excluded due to non-availability of data. In the case of Sweden, a new data source is used for 1983 and 1984. See chapter 3 n. 1 for a detailed discussion of the way to calculate the Swedish figures.

[b] It is assumed that total investment costs for robots and NCMTs is 40 per cent and NCMTs is 40 per cent and 70 per cent higher than in (3) and (5) respectively.

[c] Estimated.

[d] Budget.

[e] These are for total investment in computer-aided design and production technologies in Sweden for these years. It is thus the only information which includes investments in CAD and FMS. If we calculate the share of robots and NCMTs only, as in the case of the other countries, their share was 12.7 per cent in 1982. Other electronically controlled capital goods would then account for 5 per cent of the investment in machinery and transport equipment in 1982. This is equal to US$37.5 million.

[f] Only a minor part of the investment value in NCMTs in column (5) for the years 1983 and 1984 is multiplied by 1.4. The reason is that one source of the data is the distributors of NCMTs, who often take orders for complete units. Thus, their sales value of NCMTs is higher than the import value as reported by the customs.

[g] For all years, the data refer to investment in 'plant and machinery' and in 'vehicles, ships and aircraft'.

[h] Investment in robots for these two years was very small.

Appendix table 7.1 notes continued

There exist a number of problems with finding data on investment in machinery and equipment in the case of the USA. The data in the table are found under the heading of 'Expenditures on new plant and equipment, Major industries, Selected years, 1960 to date' and cover: 'Electrical Machinery and Equipment', 'Nonelectrical Machinery', 'Motor Vehicles and Parts' and 'Other Transportation Equipment' and are taken from NMTBA, 1985/6, p. 15. However, compared to the data in NMBTA, 1985/6, p. 18, the figure is much higher. The figures are also higher than those given by United Nations (1980). Hence, there is a risk that the data overstate the value of investments and thus under-estimate the share of flexible automation in these investments. Furthermore, the data in NMBTA, 1985/6, p. 15 include investment in plants. We have assumed that the investment in plants was 18.5 per cent of total investment which is the average for the period 1978–81 for 'Durable Goods Industries' (see NMTBA, 1985/6, p. 17). Had we used the data in NMBTA, 1985/6, p. 18 referred to above, which only cover the years to 1981, the figure for investment in machinery and equipment would have been adjusted downwards. For 1981, these data suggest an investment worth US$24.8 billion whilst the figures we have used, after adjusting for investment in plants, came to US$32.3 billion. The use of the lower figure for investment would, however, not greatly alter the figure in column (8) in the table. If, in 1984, the same proportion existed between the two sets of investment figures as in 1981, the column (8) figure would have risen to 7.0 per cent had we used the lower figure on investment.

Sources: Japan, *Japan Statistical Yearbook* (various years) for column (2). Appendix table 4.2 for column (3). Elaboration on data supplied by the Japan Machine Tool Builders' Association for column (5). Sweden, information received from the Swedish Central Bureau of Statistics for column (2). Appendix table 4.2 for column (3). Elaboration on data received from the Swedish Machine Tools Manufacturers' Association and the Swedish Association of Tool Distributors for column (5). UK, Central Statistical Office (various years for column (2). Appendix table 4.2 for column (3). Elaboration on data supplied by the British Machine Tool Trades Association for column (5). USA, National Machine Tool Builders' Association, 1985/6, p. 15, for column (2). Appendix table 4.2 for column (3). Elaboration on NMBTA, 1984/5 and 1985/6 for column (5).

8

Flexible Automation and Employment

8.1 Introduction

The relation between technical change and (un)employment has been discussed for centuries. In the present period of high unemployment in most of the OECD countries, which has occurred simultaneously with the diffusion of microelectronic-based automation techniques, e.g. flexible automation, the interest in the employment consequences of technical change has increased. In this chapter we will focus upon quantitative as well as qualitative employment aspects of automation in the engineering industry.[1] The problem of technical change and employment will be discussed at different levels – the machine level, the firm level, the national and, to some extent, the international level. We consider it crucial to pursue the analysis at these various levels of aggregation.

In section 8.2 we will discuss technical change and employment at the machine level with respect to the four techniques addressed in this book. It concludes in section 8.2.5, with an estimation of the total employment consequences of the introduction of these techniques. Section 8.3 is devoted to determinants of the level of (un)employment other than technical change, e.g. various compensation effects and political factors.

The relation between technical change and employment is a complex one. First, the employment situation in a country may be both a (partial) determinant[2] of technical change as well as a consequence thereof. Secondly, technical change is only one of many variables which determine the level of employment. It is still true, however, that the four techniques dealt with in this book are labour-saving, i.e. the amount of labour needed per unit of output is reduced with the application of these techniques. Before going into more detail about the employment consequences of flexible automation it may be appropriate to look at the trends in employment in the engineering industries of certain countries (see table 8.1). Between 1971 and 1980 employment, as measured in number of employees, decreased in three of the countries listed, whereas in the USA, engineering employment increased by 12 per cent.

Table 8.1 Employment in the engineering industry,
in thousand employees

	1971	*1975*	*1980*
USA	7,295	7,724	8,809
FRG	4,132 (1972)	3,948	4,082
Sweden	276	305	271
UK	3,596	3,364	3,100
Japan	n.a.	n.a.	5,275
Total	15,299	15,341	16,262[a]

[a] Including Japan, 21,537.
Source: ILO, 1981, table 5B; ILO, 1982, table 5B.

8.2   Technical Change and Employment at the Machine Level

8.2.1   NUMERICALLY CONTROLLED MACHINE TOOLS

To be specific about the employment consequences of flexible automation, it is necessary first to look at the problem at a disaggregated level, e.g. at the level of the machine. Let us start by looking at the employment consequences, both quantitative and qualitative, of the introduction of NCMTs. Before we do this, we should make clear what we mean by the term 'replace'. We mean a case where there are less jobs with the new techniques than there would have been with the old ones, had output expanded in the same way. A 'replacement' would therefore not necessarily mean that people have been made redundant by the new technique (Palmer et al., 1984, p. 21).

In an analysis carried out elsewhere (Palmer et al., 1984, pp. 24–5), we reached the conclusion that, on average, one NCMT substitutes for two jobs. In table 7.2, we saw that the total stock of NCMTs was estimated to be 118,157 in Japan in 1984 and 103,308 in the USA in 1983. Hence, over 236,000 jobs in Japan and 206,000 jobs in the USA had been 'replaced' through the use of NCMTs instead of conventional machine tools. This represents 2.3 per cent of the engineering labour force in the USA in 1980 and 4.5 per cent of the engineering labour force in Japan in the same year. Hence, the labour replacement effect of the introduction of NCMTs is significant.

The economic efficiency of stand-alone NCMTs is well established, and manifested through their rapid diffusion. The main source of saving arises from the increased labour productivity. They save on operators of manual machine tools, i.e. they save on skilled labour through two mechanisms. First, the large increases in labour productivity that can be attained imply that the skilled labour content per unit of output is reduced. Secondly, through the use of NCMTs, it is possible to de-skill the job of the machine tool operator. The de-skilling process is, however, not an automatic one and is certainly not exclusively determined by the new technique.

There are, in fact, many ways of using NCMTs and there are several ways in which a firm can organize the work of running NCMTs. The operator can be anything from a 'robot-like' loader and unloader of the NCMT to a very skilled operator who does the programming, setting and optimization as well as other, less qualified, tasks. However, several empirical studies suggest that, in most cases, there has in fact been a de-skilling in the operator work as compared to the use of conventional machine tools.[3] When this is the case, however, a limited number of very qualified jobs as programmers and setters are simultaneously created. In addition, more qualified personnel are needed for the maintenance and repair of the NCMTs. It seems safe to argue, however, that the amount of skilled labour per unit of output declines with the *use* of NCMTs.[4]

8.2.2 INDUSTRIAL ROBOTS

The dominant motive for investments in *industrial robots* is to decrease labour costs. The labour saved through robot investments is unskilled and semi-skilled labour. The extent of labour displacement varies considerably between applications. It is often assumed that each robot, on average, replaces two manual workers, although sometimes it is argued that five workers are replaced by one robot. In appendix table 4.1 we saw that the total number of robots installed in the OECD countries was somewhat less than 100,000 by the end of 1984. Hence, at least 200,000 jobs have so far been 'replaced' by robots in the OECD countries. In Sweden the stock of robots was about 1,900 units by the end of 1984 (see appendix table 4.1). Hence the gross replacement effect would have been 3,800 jobs with the (low) assumption that two jobs are replaced by one robot. In total there were about 271,000 employees in 1980 in the Swedish engineering industry. Thus the substitution effect is equivalent to about 1.4 per cent of the engineering labour force.

However, jobs are also *created* by robotics, mainly in four areas: robot manufacturers, suppliers to robot manufacturers, robot systems engineering and jobs within robot-using firms. Those who work in robot-using firms maintain and repair robots, while those in robot systems engineering execute the applications engineering requirements for installations of robot systems (Hunt, 1984, p. 12).

Hunt's study (1984) has estimated that for each job displaced by robotics 0.32 jobs are created in the USA in the four areas mentioned above. Thus, again in the Swedish case, using Hunt's estimate, about 1,200 jobs would have been created, leaving the net replacement to be 2,600 or less than 1 per cent of the blue collar labour force in engineering.[5]

Hunt suggests (p. 13) that 14.5 per cent of the jobs created by robotics are engineers and 38.3 are technicians, i.e. persons with training to test, program, install, troubleshoot or maintain robots. As we know, a basic precondition for the use of robots is that the work is already very repetitious and requires almost no skills, except sometimes physical strength and powers of endurance. Robots are also sometimes used in dangerous and dirty work environments. Accordingly, the jobs created and the jobs displaced through robotics do not match up very well. The jobs eliminated tend to be semi-skilled or unskilled, while more than 50 per cent of the new jobs created demand a significant technical background. This has been called the *skill twist* (Hunt, 1984, p. 14). Hence, the introduction of robots is, on the whole, positive in terms of consequences for skill requirements and the quality of jobs.

8.2.3 COMPUTER AIDED DESIGN SYSTEMS

CAD systems normally increase the productivity of designers and draughts-men, after a period during which they learn how to use the equipment. Hence, the introduction of such systems means that a smaller number of engineers are needed to undertake a given amount of design work. In other words, the main employment implication of CAD is that it saves skilled labour.

Let us, in a very speculative manner, try to guestimate how many jobs are replaced by the diffusion of CAD. In chapter 6 we saw that there were about 10,000 CAD systems installed in the world in mid-1982, i.e. before the rapid diffusion of the PC-based systems. Assuming that the number increased by 25 per cent per year, we would have about 20,000 systems by mid-1985. If each system has four workstations we would, in total, have 80,000 workstations in mainframe- and minicomputer-based systems. In addition, there were 42,000 PC-based CAD workstations. In total, this would give us about 120,000 CAD workstations globally.[6]

The labour productivity increases from the use of CAD can – as we saw in chapter 6 – vary very much depending on the specific application (new drawings, modifications, etc.). However, a conservative estimate would be 3 to 1 (see chapter 6). This ratio refers, however, to only those tasks where the designer or draughtsman uses the CAD system. Certainly these people do not use CAD during their entire working day. Assuming that they use the CAD equipment, on average, for 25 per cent of their working day, it would mean that about 90,000 design and draughting jobs have been replaced through the diffusion of CAD during this decade.

8.2.4 FLEXIBLE MANUFACTURING SYSTEMS

An FMS means, inter alia, the integration of NCMTs tools and automatic transport and handling (e.g. robots) equipment. As we saw in chapter 5, these systems may vary considerably in their size and nature, from small modules to extended systems including many machines and with a very complex workpiece flow system between them. In smaller systems only a few workers are 'replaced'. For the larger systems, there are 'stories' to be heard of FMSs where a dozen workers are needed instead of several hundred if 'traditional' production techniques were to be used, and for the production of the same volume of output. Whilst these 'stories' may well contain important elements of truth, only a very limited number of large FMSs exist, i.e. this technique has not yet reached the growth phase in an S-curve context.

Because of the widely varying nature of the FMSs and the insufficient knowledge of the stock of the various types of systems, it hardly makes sense to try to estimate how many jobs they replace. In addition, it would mean considerable double counting, since the FMSs include many NCMTs and also some robots. It is enough to stress that although FMSs are labour-saving, the order of magnitude of the total global labour replacement is 10,000 rather than 100,000 at present.[7] When they start to diffuse rapidly, they will, however, have a large labour-saving impact.

Previously in this section we have seen that NCMTs, industrial robots and CAD systems do substitute for labour, with different skill levels, but that they also create a need for new kinds of jobs – often of a more qualified nature. This mechanism is even more pronounced in the case of FMSs. It is often extremely complicated in a technical sense to coordinate the running of several different machines and to make the flow of workpieces between them function properly. Taken with the fact that this technique is in an early stage of its development, this has sometimes meant tremendous problems in implementation of the systems and in making them function efficiently. This is also linked with the supply side problems discussed in chapter 2. In firm interviews, we have learned of cases where complex FMSs did not function in an acceptable manner even after several years of intensive efforts. In one case it was even necessary to put manual drivers on fork-lift trucks which were supposed to be guided and run in an automatic manner. In another case, the entire FMS was thrown out and replaced by more traditional techniques.

Also the running and, in particular, the maintenance and repair of FMSs mean that new capabilities are needed. For example, electronics competence is self-evidently required, but such competence must be combined with mechanical engineering and hydraulics competence – preferably in the same persons. Engineers and workers with such a broad capability are extremely rare.

Hence, the technological capabilities needed to implement and run FMSs should preferably be broad and focused at the same time. This is true also for the other automation techniques, but in a less pronounced way. We believe that lack of such capabilities is producing a very important bottleneck in the further diffusion of flexible automation techniques, particularly FMSs. In order to support the diffusion of these techniques, it is therefore extremely important that the education system be adapted to create future employees with the capabilities needed. This is a concern for the general (infra-structural) education system and should therefore be a matter of state policy. But it also relates to internal educational efforts within the firms supplying and running these techniques.

8.2.5 TOTAL EMPLOYMENT EFFECTS AT THE MACHINE LEVEL

Let us try to summarize the quantitative employment effects of flexible automation in the engineering industries of the OECD countries. In table 8.2 we have calculated the labour-saving consequences of the diffusion of NCMTs and industrial robots and related these to total employment in the same sector for the USA, Japan, Sweden, the UK and the Federal Republic of Germany. The table suggests that 3.7 per cent of the engineering labour force has been 'replaced' by robots and NCMTs in the USA, Japan, Sweden, the UK and FRG taken together. The percentage varies between 2.3 for the UK and 6.9 for Japan.

It must be underlined that whilst the estimate in the table is somewhat rough, it is also very conservative. Let us elaborate on this. First, we have assumed that each robot replaces only two workers although a common range discussed is two to five. One author even uses the range two to seven workers displaced by each robot (Ebel, 1986). Secondly, the jobs displaced through the use of CAD systems and FMSs are not included.

Thirdly, the discussion in this section has mainly concerned those jobs that are *directly* related to the actual implementation and running of the flexible automation techniques. However, flexible automation also affects employment with regard to 'peripheral' functions. The new techniques often need less space than those replaced. Hence fewer construction workers and a smaller number of cleaning personnel are needed. The administration of production also often becomes less complex and less labour-using. Indeed, one of the main savings achieved by Flygt when they implemented their new workshop, described in section 7.2.1, was in the form of reduced need for white collar workers, e.g. production planners. These examples could be

Table 8.2 Number of jobs 'replaced' in engineering industries by NCMTs and robots in certain OECD countries

	NCMTs (1984)	Robots (1984)	Total replaced	Total employees (1980, 1000s)	No. replaced as % of total employees
USA					
stock	103,308[a]	13,000			
replaced	206,616	26,000	232,616	8,809	2.6
Japan					
stock	118,157	64,657			
replaced	236,314	129,314	365,628	5,275	6.9
Sweden					
stock	6,010	1,900			
replaced	12,020	3,800	25,820	271	5.8
UK					
stock	32,566	2,623			
replaced	65,132	5,246	70,378	3,100	2.3
FRG					
stock	46,435	6,600			
replaced	92,870	13,200	106,070	4,082	2.6
Total					
stock	306,476	88,780			
replaced	612,952	177,560	750,512	21,539	3.7

[a] 1983.
[b] For both techniques we have assumed that one unit replaces two jobs.
Sources: Table 7.2 for stock of NCMTs. Appendix table 4.1 for stock of robots. Table 8.1 for employment data.

multiplied, but the main tendency of these changes is that the labour-saving character of the new techniques in the engineering industry is further pronounced when the process of production as a whole is taken into account. However, it is not possible here to gauge these additional, peripheral employment consequences. Finally, it must be mentioned that flexible automation is certainly not the only technical change currently taking place in the engineering industry.

The obvious conclusion is that the saving of labour achieved through the use of flexible automation in the engineering industry is very significant and can by no means be ignored in discussions about what determines the level of (un)employment in an economy.

8.3   Other Determinants of the Level of (Un)employment

The fact that new flexible automation techniques are labour-saving at the level of the machine does not, however, necessarily imply that their implementation leads to higher unemployment in the society as whole. Their introduction may even be combined with a lower rate of unemployment. Indeed, it may be noted (see table 8.2) that the displacement percentages are highest for Sweden and Japan, which are the two countries having the *lowest* rates of unemployment of the countries included in the table.[8] The reason is of course that the level of (un)employment in a society is determined by many other variables, as we mentioned at the beginning of this chapter. New techniques have been labour-saving during the whole history of technical change. These labour-saving consequences are often compensated for by changes in other variables, e.g. increased competitiveness leading to a rise in demand for labour, or government policy leading, for example, to expansion of employment in other sectors of the economy. These issues were dealt with by Palmer, Edquist and Jacobsson (1984) in a study upon which the following sections are partly based.

8.3.1   THE FIRM LEVEL

We have seen above that the various flexible automation techniques are labour-saving in that less labour is required per unit of output. This is at the machine level. However, at the level of the firm, there may be no fewer people employed, i.e. there can be substantial compensatory effects due to the increased competitiveness of the firm or country or from a general exogenously determined increase in demand. Demand may influence output and what happens to output is crucial. If output expands sufficiently, it can (more than) make up for the loss of employment due to the increases in labour productivity. It is often assumed that the growth in output will not be sufficient, but it is often an implicit assumption and is not made clear (Cooper and Clark, 1982, p. 7).

Compensatory effects may be seen in the case of Sweden. Although there was a general decrease in the number of hours worked, Elsässer and Lindvall's study (1984) of 136 firms in the machinery industry showed no relationship between increases in productivity and a reduction in the number of hours worked. A strong positive correlation was found between changes in output and labour productivity. In other words, these firms may have succeeded in obtaining a greater volume of work, although the correlation could also reflect better capacity utilization.

Thus, not even the step from labour 'replacement' at the machine level to a reduction in the employment level of the firm is a simple one. It is likewise problematical to generalize replacement effects at the machine level to the sector level in one country because of compensation effects. A specific country may be particularly competitive or uncompetitive due to technical change (or lack of), and/or the sale of this country's products may be helped by devaluation or hindered by an over-valuation of the currency. Government policy, which we will deal with in more detail below, can also be extremely influential in affecting the level of demand of the economy.

It may, however, be more feasible to generalize from the productivity increases at the level of the firm to worldwide employment in a sector, e.g. the engineering sector. Then, the country-level compensation effects due to an increased share of the world market would be eliminated.

8.3.2 THE NATIONAL LEVEL

For an individual country, an extremely important determinant for the level of employment is government policy. First, the government can directly employ people. Secondly, the government can indirectly affect employment levels by, for example, influencing investment and consumer demand. Both these factors are important in determining output and thereby employment.

Both investment and consumer demand, but most importantly the former, can be affected by government policy. Government policy can have a powerful effect on businessmen's expectations about the future of their business and hence affect their propensity to invest. Is the government pursuing an expansionary or deflationary economic policy, for example, and what are its stated plans about the economy? Other government policies could be to create incentives to invest as well as, for example, to provide help to evaluate technologies.

Government policy on employment depends on the government's general social and political aims – i.e., how important employment is considered to be in comparison with other aims. At the moment, the governments of many countries consider that reducing inflation is of greater importance than trying to maintain existing employment levels or reducing high unemployment. Here it is of interest to compare Swedish and British government policy in this area, in the mid-1980s. Low unemployment is an important aim in Sweden, whereas in Britain, unemployment has been allowed to soar in the interest of reducing inflation, which is considered by the government to be of greater importance.

Both Sweden and Britain are, of course, capitalist countries and have open economies. This means that their governments' freedom of action is more limited in comparison with socialist countries where the government has a freer hand with regard to planning employment and income distribution. Socialist countries can, for example, choose whether a given product is to be imported or whether it is to be produced at home.

In this context, an example illustrating the relationship between technical change and employment in a socialist country may be illuminating. In Cuba after the revolution (January 1959) the development strategy and political priorities with regard to employment changed drastically. As can be seen from table 8.3, unemployment declined rapidly only a few years after the revolutionary takeover. It decreased from 13.6 per cent in 1959 to 1.3 per cent in 1970 in spite of the fact that about 50,000 manual workers were replaced by mechanical sugar cane loaders between 1964 and 1970 (Edquist, 1982, ch. 5; Edquist, 1985b, p. 148). Since the possibilities for trading did not improve during the 1960s, the change in unemployment levels must be attributed to the new government's development strategy. In particular, the structure of the economy was radically transformed; between 1958/1959 and 1970 for example, employment increased by 50 per cent in industrial activities, by 90 per cent in construction, by 100 per cent in transportation and communication and by 23 per cent in services (Edquist, 1985b, p. 99). Furthermore, the pension age was lowered to 60 for men and 55 for women. In reservation, we must mention that part of the reduction in overt unemployment was due to disguised unemployment (Edquist, 1985b, p. 34).

In table 8.3 we can also see that unemployment rose during the 1970s from 1.3 per cent at the beginning of the decade to 4.1 per cent in 1980. This coincides with a drop in employment in sugar cane harvesting of between 120,000 and 160,000 owing to mechanization. The machette was

Table 8.3 Open unemployment (annual average, per cent) in Cuba: selected years 1943–81

1943[a]	21.1	1967	5.3
1953[b]	8.4	1968	4.3
1956[a]	20.7	1969	2.9
1957[b]	9.1	1970	1.3
1956–7	16.4	1971	2.1
1957	12.4	1972	2.8
1958	11.8	1973	3.4
1959	13.6	1974	3.9
1960	11.8	1975	4.5
1961	10.3	1976	4.8
1962	9.0	1977	5.1
1963	8.1	1978	5.3
1964	7.5	1979	5.4
1965	6.5	1980	4.1
1966	6.2	1981	3.4

[a] During non sugar cane harvest season.
[b] During sugar cane harvest season.
Sources: Edquist, 1983, p. 43; Edquist, 1985b, p. 24; Brundenius, 1984, table 1.

replaced by combine harvesters. This drop in employment was equal to between 4.6 and 6 per cent of the total Cuban workforce in 1970. As a proportion of the agricultural labour force it was 15–20 per cent. Superficially one might say that 120,000–160,000 people were made redundant by sugar cane harvesters.[9]

On closer examination, however, one finds that the increase in unemployment in the 1970s coincides with the entry into the workforce of large numbers of women – there was in fact a doubling of the female workforce from 482,000 in 1970 to 1,108,000 in 1980 (Brundenius, 1983, table 3), and in the period in question, the workforce increased faster than population growth. Furthermore, women have a much higher rate of unemployment than men – men 2.5 per cent in 1979 and 1980 as compared with 1.3 per cent in 1970 while for women the figure was 12.0 per cent and 7.8 per cent, as compared with 1.2 per cent in 1970 (1983, table 4). As women have not worked as manual cane cutters, and the increase in unemployment is largely women's unemployment, we cannot explain the unemployment in Cuba as being directly caused by the introduction of sugar cane harvesters. In the same period (the 1970s), the labour force was employed through the *expansion* of other industrial sectors. Between 1970 and 1979 employment in manufacturing and mining increased from 533,000 to 652,000, i.e. by 119,000 jobs. In construction, it increased from 157,000 to 256,000, i.e. by 99,000. In services the increase was from 622,000 to 934,000. Hence 312,000 new service jobs were created, many of them in health and education (Brundenius, 1984, table A1.1).

It is important to differentiate between need and effective demand. Before the revolution in Cuba there was a great need for education and health care but it was too expensive for most people. At the same time many people were unemployed – an idle or wasted resource. Thus, there was a need for services, but no effective demand because of the cost of the services and also because of lack of income caused by unemployment. After the revolution the price of these services was greatly reduced and the demand for them rose dramatically. People were then eventually able to be employed in these sectors after appropriate training. Unfulfilled basic human needs had then been satisfied through employment of previously idle resources which, in turn, made the expansion of the service sector possible.

The Cuban example shows how important political priorities can be for employment levels. It also shows that it is possible to have a fairly low degree of unemployment despite considerable mechanization in agriculture and the entry of many women into the workforce. Finally, it shows that a structural change in the organization of society can have substantial employment effects.

We would stress that structural changes are perfectly normal to all economies and can certainly occur without a social revolution. There are incremental structural changes in all economies to varying degrees. Some examples are the changes in the ownership structure in many industrialized countries. Pension funds, insurance companies and various foundations own company shares to increasing degrees in many countries. In the case of Sweden, wage earners' funds have recently been introduced. They are made up of appropriated profits from firms which will be used to buy shares on the stock market. In the long run, the sum of various marginal structural changes may well considerably change the structure of the society and may well have substantial employment effects.

The conclusion is that the unemployment level in a country is not primarily determined by the introduction of new techniques, although they have some bearing on employment, particularly on the relative employment levels in the various sectors of the economy. Although the labour 'replacement' effects of the introduction of flexible automation techniques are quite substantial, it is impossible to equate the aggregated effects of technical change at the micro level to the unemployment level in the economy as a whole. Factors at the macro level influence greatly the overall level of unemployment. Structural changes in the society as well as government goals and policies are extremely important for the level of unemployment. In the final analysis, therefore, the level of unemployment is more a social and political problem than an economic one. It is certainly not primarily a matter of the choice of techniques (Palmer et al., 1984, pp. 54–5). At the same time, union resistance to technical change tends to be strong if the rate of unemployment is high. Therefore, if a government – or some other actor – wants to raise productivity in a country by introducing labour-saving techniques, the most efficient way to avoid union resistance is obviously to secure that the general rate of unemployment is low.

Notes

1 The chapter is partly based upon a study by Palmer, Edquist and Jacobsson (1984).
2 Unemployment may – through complex social mechanisms – be an obstacle to the adoption of labour-saving techniques and a labour shortage may be a driving force behind such adoption (Edquist, 1985a; 1985b, pp. 14, 75–106).
3 See Palmer et al. (1984) for the basis of this argument.
4 The average level of skill requirements of all those involved in production, installment, programming, setting, optimization, operation, maintenance and repair of NCMTs may very well increase at the same time. Compare section 3.4.
5 The job displacement effect may, of course, be more significant in a given occupation, industry or geographical area.
6 This means that we assume a growth of 44 per cent, which is slightly below the assumed growth rates for some countries in table 7.2.

7 As was noted in chapter 5, there are around 340 FMSs installed worldwide. If each of these 'replaces' 30 workers, roughly 10,000 workers would have been 'replaced' worldwide by these large FMSs.

8 In 1985 the official rates of unemployment for the countries included in table 8.2 were as follows: USA, 7.2 per cent; Japan, 2.6; Sweden, 2.8; UK, 11.9; FRG, 8.3 (Source: OECD, 1986, tables 17, 18). In the text, labour displacement because of flexible automation in *one* sector of production (the engineering industry) is related to levels of unemployment in the economies as a *whole*. However, there are indications that Japan and Sweden are experiencing a more rapid diffusion of labour-saving techniques than the other countries included in table 8.2 in other parts of the economy as well.

9 The labour-saving due to the diffusion of sugar cane combine harvesters represented a dramatic decrease in agricultural employment and a drastic change in the structure of employment in Cuba as a whole. As a matter of fact, it was probably the largest change in this respect in Latin America this century.

Part III

The Diffusion of Flexible Automation Techniques in the Engineering Industries of the Newly Industrializing Countries

9

Numerically Controlled
Machine Tools

In this chapter we will describe the diffusion of NCMTs in the newly industrializing countries (NICs) and also discuss the determinants of and obstacles to this diffusion. Section 9.1 will give a short description of the rate and level of diffusion of NCMTs as well as their distribution between firms of different size. In section 9.2 we will discuss the determinants of the actual diffusion and the obstacles to a further diffusion of NCMTs in the NICs. The discussion of determinants and obstacles will be conducted within the conceptual framework presented in chapter 2.

9.1 Rate and pattern of diffusion

9.1.1 RATE OF DIFFUSION

NCMTs are clearly being diffused in the NICs as well as in the developed countries. In table 9.1 the estimated *stock* of NCMTs is given for a number of NICs. We can see that Korea is the largest single user of NCMTs, with a stock of 2,680 units in 1985, followed by Brazil (1,711) and India (1,178). This stock of NCMTs is growing at a reasonably fast rate, as can be seen in appendix table 9.1. In the case of India, the average cumulative growth rate was nearly 40 per cent in the period 1980–5 whilst that of Korea was 22 per cent. Whilst these growth rates are quite impressive, they are nevertheless not much higher than those of the developed countries in the same period.

As is seen in appendix table 9.2, there is a fairly steady growth of 11–18 per cent in the stock of NCMTs in both the UK and the USA. Although the growth in the number of NCMTs is higher in some NICs for some years, the growth rates in the NICs would appear to be too low in relation to those that would be needed if these countries were to 'catch up' with the OECD

Table 9.1   The estimated stock of NCMTs (units) in a number of NICs

Argentina (1985)	500
Brazil (1985)	1,711
India (1985)	1,178
Korea (1985)	2,680[a]
Mexico (1984)	500
Singapore (1985)	700
Yugoslavia (1983)	1,232

[a] We exclude the 258 NCMTs imported in 1976 with very low unit value, but add an estimated 200 machining centres and NC metal-forming tools excluded in the official import data of Korea (see Edquist and Jacobsson, 1985a, pp. 649–50). Edquist and Jacobsson, 1985, is used for data on the flow of NCMTs in the period 1973–82. Korea Machine Tool Manufacturers' Association (KMTMA), 1986, is used for the period 1983–5.

Sources: Argentina and Brazil, Chudnovsky, 1986. India, Central Machine Tools Institute, 1986; Machine Tool Census. Korea; Edquist and Jacobsson, 1985a, and KMTMA, 1986. Mexico; Mercado, 1984. Singapore; Economic Development Board, 1986. Yugoslavia, UNCTAD TT/67.

countries within a reasonable time period. Indeed, for Argentina, Brazil, India and Korea taken jointly, the cumulative growth rate in the stock of NCMTs was only 23 per cent in the period 1980–5.

However, the growth rate of the stock of NCMTs is not a good indicator of the extent to which a country invests in order to benefit from the new technique. The growth rate of the stock of NCMTs is clearly influenced by the overall rate of investment in machine tools, which indeed may be very high in some of the NICs. For example, in Korea the ratio between investment in machine tools and the value added in the engineering sector was 9.2 per cent in 1981 (National Machine Tool Builders' Association, United Nations, 1982) whilst in the USA, UK and Sweden it was only 1.4, 1.9 and 1.3 per cent respectively in 1979 (NMTBA, 1983/4; United Nations, 1980). Hence machine tools investment in Korea is relatively much higher than in the OECD countries mentioned. Even if the share of NCMTs in total machine tool investment is low, the growth rate in the stock of NCMTs in Korea may be high. Hence the growth rate in the stock of NCMTs is not a good indicator of the propensity to invest in NCMTs instead of conventional tools.

The rate of diffusion of NCMTs is more appropriately indicated by measuring the share of investments in NCMTs in total machine tool investment. This indicator shows more clearly the extent to which NCMTs are chosen instead of conventional machine tools. In table 9.2 it can be seen that this share ranges between 7 and 23 per cent for those NICs for which we have data. For the OECD countries, the share ranged between 40 and

62 per cent in 1984 (see table 3.2).[1] Hence, the NICs are far behind the OECD countries in terms of the extent to which they choose NCMTs instead of conventional machine tools. This also applies to Korea, which is the NIC with the largest stock of NCMTs and which shows the greatest intensity in use of NCMTs as measured by share of total machine investment.

Table 9.2 Estimated yearly investment in NCMTs in relation to total investment in machine tools in some NICs

(1)	(2)	(3)	(4)	(5)	(6)
		Investment in NCMTs		Investment in	
Country	Year	Units	Value (US$m)	machine tools (US$m)	(4)/(5)×100 (%)
Brazil	1982	150	26.2[a]	236	11.1
India[b]	1984	188	39.8	306	13.0
Korea	1982	178	12.4[c]	136	9.1
	1983	353	30.7[c]	235	13.1
	1984	458	59.8[d]	257	23.2
	1985	593	58.3[e]	388	15.0
Yugoslavia	1982	120	21.0[a]	288	7.3

[a] A unit price of US$175,000 is assumed.
[b] The Indian data on investment in NCMTs refer to the NCMTs that were given an import licence in 1984. All of these may not have been imported. On the other hand, we exclude local production, which was in the order of 10–20 units in 1984.
[c] For 1982–3 the figure includes investment in NC lathes, NC milling, NC drilling and NC boring machines. In addition, Japanese exports to Korea of machining centres are included. We have assumed that total import of machining centres is double that of imports from Japan.
[d] NC metal-forming machine tools, NC grinding, honing and lapping machines are excluded.
[e] Only NC lathes, NC milling machines and machining centres are included. We assumed an exchange rate of 850 won/US$.
Sources: Brazil, Rattner, 1984; National Machine Tool Builders' Association, 1983/4. India, Directorate General of Technical Development, 1986; NMTBA, 1985/6. Korea, Korean Machine Tool Manufacturers' Association, 1985, 1986; Office of Custom Administration, foreign trade statistics, 1983, 1984. Yugoslavia, UNCTAD TT/67; NMTBA, 1983/4.

9.1.2 DISTRIBUTION OF NCMTs
ACCORDING TO THE SIZE OF FIRMS

Data on the size distribution of the users of NCMTs vary in type and quality. In Argentina data collected in 1981 suggested that only 25 per cent of the stock of NCMTs were in firms that had more than six units. Furthermore, there were many small machinery and component producers or subcontractors with only one NCMT (Jacobsson, 1986, ch. 6). Chudnovsky also suggests that '...there are a few users having a relatively large stock, say of eight to ten units or more, but most of them have around two or three units' (1984, p. 17). Chudnovsky furthermore says that '...most of them [NCMT users] are medium-sized firms. A few small, and even very small firms have started to rely on this equipment' (p. 17).

To the extent that 25 per cent of the stock is held by a very small number of large firms and that the remaining users have between two and three units on average, the total number of users would amount to around 100, give or take 20 per cent.

In Brazil, Rattner (1984, p. 8) claims that 4 per cent of the users have (it is not clear for which year) nearly half of the installed stock of NCMTs. In 1979 or 1980, furthermore, the use of NCMTs was largely concentrated to large firms. According to Tauile (1984, p. 61), 42 per cent of the users had more than 1,000 employees; 24 per cent between 500 and 1,000; 22 per cent between 100 and 500 employees and only 7 per cent less than 100 employees. The total number of users was 157 in 1980 and estimated to be about 200 in 1984 (Rattner, 1984, pp. 6, 12).

In a survey we undertook in 1983 on the diffusion of NCMTs in India, sales data for firms that had installed 240 NCMTs were available. These firms accounted for roughly two-thirds of the production value in the formal engineering sector in 1979. We define a small firm as one having a sales value of less than US$7 million.[2] In terms of value of output this corresponds broadly to a US firm with 100 employees, which was defined as a small firm in chapter 3. Only six of the installed NCMTs were in such firms. Another 33 NCMTs were installed in firms with a sales value of between US$7 and 32 million which would, very roughly, be equivalent to a US firm with between 100 and 500 employees.[3] The remaining NCMTs (of the 240), i.e. the vast majority, were installed in large firms. These data are in line with those of the Central Machine Tools Institute (1986), who suggest that only 6 per cent of the stock of NCMTs in 1985 was in 'small scale units'.

Finally, in the case of Yugoslavia UNCTAD (TT/67, p. 15) reports that 'More than one third of the NCMTs is installed in five top ranked firms, another third is accounted for by...20 firms..., and less than a third of the NCMTs is installed in more than 100 remaining firms, mostly medium-sized and some small firms'. Thus, with the possible exception of Argentina, larger firms in the NICs dominate the use of NCMTs to an extent that resembles the diffusion process of the new techniques in the OECD countries at a much earlier time.

In conclusion then, whilst NCMTs are being diffused in the NICs to a considerable extent, the share of NCMTs in total investment in machine tools is very much lower than in the leading OECD countries. Furthermore, with the exception of Argentina, it appears as if it is mainly the larger firms which adopt NCMTs.

9.2 Determinants of Diffusion of Numerically Controlled Machine Tools

9.2.1 INTRODUCTION

As indicated in chapter 2, the determinants of the diffusion of new techniques are manifold and the relation between them is complex and varies between cases. In the case of NCMTs, a basic precondition for diffusion is self-evidently that there is a demand for machine tools in the country in question. In other words, it must have an engineering industry. The structure of the engineering industry is also important, since NCMTs (and conventional machine tools) are used with different intensities in the industry's subsectors (see chapter 3). Provided that machine tools are demanded at all in the economy, the choice between various kinds of machine tools – e.g. conventional and numerically controlled ones – becomes a relevant issue.

For the given reasons above, in this section, we will first briefly address the issue of the size and structure of the engineering industry (ISIC 38) in the NICs. Thereafter we will proceed to a discussion of the more specific problem of diffusion of NCMTs, which is then seen mainly as a choice between these kinds of machine tools and conventional ones.

9.2.2 THE SIZE AND STRUCTURE OF THE ENGINEERING INDUSTRY

The engineering sector is relatively advanced in all the NICs in that it constitutes a sizeable share of manufacturing value added. In appendix table 9.3 it can be seen how this share ranges from 21 per cent in Mexico in 1979 to 50 per cent in Singapore in 1980. As a comparison, we have also included the cases of Sweden and the UK, which had a share of 43 and 39 per cent respectively.

In terms of the absolute size of the engineering sectors in the NICs, it can generally be said that value is small, indeed very small in some countries. The exception is Brazil which, however, also includes metallurgy in its statistics. Adding the engineering sector value added of the developing countries (including Yugoslavia) in appendix table 9.3 gives a figure of US$37 billion, which is only 2.8 times that of Sweden which has only 8 million inhabitants. Thus, although the engineering sector is relatively well developed in these countries, and therefore one of the basic conditions for a diffusion of NCMTs does exist, the absolute size of the engineering industries in these economies taken jointly is rather small.

The use of NCMTs varies greatly between sectors within the engineering industry. The bulk of the NCMTs are – in Japan and the USA – used within ISIC 382, which is the non-electrical machinery industry (see table 3.4).

The transport machinery industry (ISIC 384) is also a large user of NCMTs in these countries. Within the latter, the bulk of NCMTs are used within the automobile components and aircraft industries. In table 9.3 we can see the structure of the various engineering industries in the NICs at three digit ISIC level. ISIC 382 is the relatively strongest category in Brazil and India, where the share of ISIC 382 in total value added is at the same level as in Sweden and the UK. Korea, Mexico, Argentina and Singapore are in this respect structurally the weakest NICs.[4] ISIC 384 is also fairly important in several of the NICs. Automobile production (including components) is developed in most of the NICs.

Table 9.3 Structure of the engineering sector (% of value added) in some NICs, Sweden and the UK

		381	382	383	384	385
Argentina	(1973)	23	20	16	39	2
Brazil	(1979)	35[a]	29	8	18	n.a.
India	(1978)	11	31	27	28	2
Korea	(1982)	16	14	35	32	3
Mexico	(1979)	26	16	18	40	n.a.
Singapore	(1981)	11	23	40	24	2
Yugoslavia	(1980)	30	26	23	20	1
Sweden	(1979)	18	30	19	31	2
UK	(1979)	19	29	20	28	5

[a] Including metallurgy.
Sources: Korea, Economic Planning Board, 1982; Yugoslavia, India and Mexico, United Nations, 1980; Argentina, Sweden and UK, Jacobsson, 1986; Brazil, Annuario estadistico do Brasil, 1982.

9.2.3 DISCUSSION OF DETERMINANTS

9.2.3.1 Three Approaches Reconsidered The choice of technique is a very complex issue. Economic, technical, institutional and political considerations influence the outcome, as well as such factors as the prestige gained from having new techniques in the firm. In the discussion below of choice between conventional and numerically controlled machine tools various issues and approaches addressed in chapter 2 will be relevant. In section 2.2 we addressed three approaches, now summarized below.

1 In the first approach we mentioned three categories of factors that influence the diffusion of new techniques. The first two categories are important because of their effects on the profitability of the new technique. The three categories were: (i) the character of the technique; (ii) the nature

of the structural environment of the actors; and (iii) the characteristics of the actors as such.

(i) The nature of a new technique, in relation to the old, influences the profitability of adopting the new technique. The main characteristics of NCMTs are that they save labour, particularly that of skilled machine tool operators, compared to conventional machine tools.

(ii) With regard to the structural environment of the social actor making the choice between different techniques, factor endowments and factor price relations are important determinants of the diffusion of NCMTs. This environment also includes infrastructural conditions such as educational level, and credit availability as well as institutional factors like government policy with regard to industry and trade and legal conditions. In other words, it is necessary to introduce several institutional and political factors into the analysis.

(iii) Thirdly, the character of the actors as such can be expected to influence the choice of technique. For example, in terms of the propensity to take risks, it matters whether the firms potentially introducing NCMTs are large or small. Whether they have close foreign relations or not can affect the firms' information about new techniques. The attitude of a firm's management may also be important. Attitudes may differ with regard to the risks involved in introducing NCMTs. Some managements may also want to invest in new techniques for status reasons or to learn more about advanced manufacturing techniques and thereby raise the technological level in the firm.

2 A dynamic way of looking at the diffusion of techniques is to do so within the framework of the S-curve concept, illustrating the pattern of diffusion of a new technique by dividing it into three phases: introductory, growth and maturity. The determinants behind the movement of a product along its S-curve are quite complex. Of particular importance in the introductory as well as the growth phases are supply side considerations, e.g. the nature and behaviour of the suppliers of the technique and their diffusion of information and knowledge about the technique.

3 The third approach presented in section 2.2 was expressed in terms of the concept 'social carriers of techniques'. This concept was defined in terms of six conditions: (1) *interest*, (2) *organization*, (3) *power*, (4) *information*, (5) *access* and (6) *knowledge*. All these conditions must be fulfilled for a certain social actor with regard to a specific technique – e.g. NCMTs – in order for implementation to take place. As a corollary this means that those of the six conditions which are unfulfilled constitute obstacles to diffusion and thereby points of intervention in a policy of supporting the diffusion of new techniques. Many of the determinants included in the previous approaches are dealt with also in this approach, but in a more general form.

In the following sections we will discuss the determinants of the diffusion of NCMTs in the NICs in terms of a mixture of the approaches. The more general approach of social carrier of techniques will provide the overall structure of the discussion, whilst the other approaches will be used in the analysis of more specific determinants.

As we saw in section 2.2.3, conditions 2 (organization) and 3 (power) of the six conditions defining a social carrier of techniques are normally automatically fulfilled by firms in capitalist countries with regard to NCMTs. Therefore we will refer mainly to condition (1) (interest) and to conditions (4) to (6) (information, access and knowledge) in our discussion of determinants of and obstacles to diffusion of NCMTs.

The analysis of condition (1), i.e. the interest of firms for introducing NCMTs will draw on the first theoretical approach, presented in section 2.2.1, i.e. the first approach listed in this section. Necessary conditions for firms to introduce NCMTs, however, are that they have information about, access to and knowledge of how to implement and use NCMTs. Let us therefore start by discussing these prerequisites and discuss the interest condition afterwards.

9.2.3.2 Information A necessary condition for the diffusion of a new technique in an economy is of course that potential adopters have information about its existence and that it can be advantageously used within a particular economy. The firms must have information about the benefits that an NCMT can give. As has been shown in the literature, the differences between firms with regard to awareness of a new technique can amount to many years (Gebhardt and Hatzold, 1974, pp. 42–3). The diffusion of information concerning the availability of and the advantages of using a new technique can thus be very slow and uneven. Thirty years after the first NCMT was produced, information about NCMTs is still very restricted among many firms in the newly industrializing countries. Let us take, as an example, Yugoslavia:

> As a consequence of a general delay in diffusion of information, such machines are quite often purchased for an inappropriate type of manufacturing. Overall information on the possibilities of these technologies is, on average, at a low level. For instance, a good number of firms do not even know that NCMTs are very appropriate for small series of production. (UNCTAD TT/67, p. 47.)

In the very early stages of diffusion of a new technique in a developing country information about it is scarce in the economy. Local suppliers do not exist and distributors of foreign made machines may not be represented in the country. If they are, they may very well not put a great deal of

emphasis on marketing etc. This was the case in Argentina in the mid 1970s
as regards suppliers of numerical control units (Jacobsson, 1986, ch. 6).
Information about the new technique would then be expected to be concentrated among larger firms who have some kind of connection with foreign
firms either through equity and/or through licensing arrangements. Thus,
the large firms would be expected to adopt the new techniques earlier than
smaller firms. This is also generally the case in the NICs. Rattner (1984,
p. 5) describes how NCMTs were first diffused in Brazil. Ford bought the
first NCMT in 1967–8 followed by Worthington in 1969 and Clark in 1970.
Tauile (1984, p. 60) also suggests that in 1980, 62 per cent of the stock
of NCMTs was installed in firms with foreign capital. As a comparison,
in 1979 foreign firms' share of the equity amounted to 36.5 per cent in the
machinery industry, 37.5 per cent in the electrical and communication
industry and 57.2 per cent of the transport equipment industry (World
Bank, 1982, p. 13).

In the case of Argentina, we saw in section 9.1 that NCMTs are now
being diffused also to smaller and medium-sized firms. These are probably
locally owned. The first user of NCMTs in Argentina is, however, said to
have been the local subsidiary of the US company Hughs Tool, which is
one of the most important users of NCMTs in Argentina today. The first
NCMT was bought in 1970 by this firm and it had 33 units in 1983. In
India we saw in section 9.1 that larger firms still dominate the market for
NCMTs. One of the pioneers in this field was Hindustan Aeronautics, which
was the largest user of NCMTs in India in 1983. Other early users in India
were the large state-owned Bharat Electronics Company and Bharat Heavy
Electrical Company (authors' own survey).

Of course, there are reasons other than connections with foreign firms
and easy access to information why larger firms would be expected to adopt
NCMTs earlier than smaller ones. Larger firms can normally take the risks
of investing in an unproved technology more easily than smaller firms. Yet
another reason why these larger firms choose NCMTs earlier may be the
types of products that they produce. At least some of the major users of
NCMTs in the NICs produce very complex products, such as aeroplanes,
oil drilling equipment and nuclear power plants. NCMTs may be chosen
for reasons of securing an adequate and even level of precision. This is an
issue to which we will return below.

9.2.3.3 Access A central feature of the global NCMT industry is the
wide differentiation of its products. CNC lathes or machining centres are
sold in very many models and sizes, with greatly different performances and
with different degrees of standardization. No local industry, with the
possible exception of the Japanese, can or does supply users with all types
of NCMTs.

In terms of sources of supply of NCMTs in the NICs, we can see the picture in Argentina, Brazil and Korea in table 9.4. From a very high import dependence at the end of the 1970s, the local industry (which in Brazil includes foreign transnational corporations) in these countries now supplies over half of the delivered NCMTs. In the case of India, imported NCMTs account for 88 per cent of the stock (Central Machine Tools Institute, 1986). Given, however, the large number of entries into this industry recently (Edquist and Jacobsson, 1985a), we would expect that this ratio will decline considerably in the near future. As a comparison, the import share of investment in the UK in 1984 was 50 per cent and the corresponding US figure for 1983 was 52 per cent (elaboration on Machine Tool Trades Association and Metalworking Production, 1985 and National Machine Tool Builders' Association, 1984/5). Hence, the import share of investment in these OECD countries was higher than in Argentina, Brazil and Korea.

Table 9.4 Import share of investment in NCMTs in Argentina, Brazil and Korea, 1978–83 (% of units installed per year)

Year	Argentina	Brazil	Korea
1978	100	79	n.a.
1979	98	63	99
1980	97	31	93
1981	85	44	71
1982	61	20	23
1983	44	n.a.	45

Sources: Argentina, Chudnovsky, 1984, table 3.1; table 9.1 above. Brazil, Rattner, 1984, tables 1 and 3. Korea, Edquist and Jacobsson, 1985; Korea Machine Tool Manufacturers' Association, 1985; Office of Custom Administration, monthly foreign trade statistics, 1984.

Obviously, this may mean that the users in the NICs may have to be satisfied with a more narrow choice of sizes and models than their European or US counterparts, which in turn may mean that the user opts less frequently for the choice of NCMTs rather than conventional machine tools. Thus *access* to some versions of the new technique may be limited for the NICs' firms.

9.2.3.4 Knowledge The use of NCMTs as opposed to conventional machine tools normally means a major change in the complexity of the technique used. Furthermore, the use of NCMTs implies that for the bulk of the buyers an essential part of the technique – i.e. the electronic one – is in the form of a 'black box'. A conventional lathe, for example, is essentially a simple technique and a skilled craftsman can master the maintenance of the lathe fairly easily. This is not the case with NCMTs where *knowledge* of the technique is embodied in specialist skills which need to be developed in order to maintain the NCMT. In larger firms, in-house maintenance teams can and are normally developed. In Argentina, for example, it is reported that larger firms send personnel abroad to learn the basics of maintaining the NCMTs. In one case, in a firm having 23 NCMTs, the maintenance group consisted of some eight to ten engineers and technicians with an education in electronics (UNCTAD TT/66, p. 30).

In contrast to the larger firms that use NCMTs smaller firms in Argentina have to rely much more on the supplier of the NCMTs for both the training of programmers and to ensure an adequate maintenance service. According to UNCTAD TT/66 (p. 32), however, the introduction of NCMTs into smaller firms seems not to have been associated with any major difficulty as regards training or maintenance. It should be noted, however, that Argentina now has a sizeable stock of NCMTs installed. In earlier years there were problems with the supply of skills and services as regards the teaching of the essentials of the new technique and the repair and maintenance of the control unit (Jacobsson, 1986, ch. 6). Clearly, in such a situation it is easier for a larger firm to build up an in-house knowledge about how to repair and maintain NCMTs.

In Yugoslavia there was, at least as late as in 1983, a clear problem of supply of maintenance services and a general lack of maintenance skills:

> According to the standards [planned time for maintenance and damages] actual time spent for maintenance and repairs was more than two times longer than it supposed to be [*sic*]. The main reason for such a situation is insufficient mastery of NCMTs, a lack of spare parts due to import restrictions, and the lack of maintenance skill. This is particularly true for the general machinery branch, with the largest number of medium size and small enterprises where maintenance and damages came to 45 per cent of all standstills in 1983. (UNCTAD TT/67, p. 29.)

Thus, it is the small and medium-sized firms which face the greatest problems with maintenance in Yugoslavia also. Apart from the reasons suggested in the quotation above, the UNCTAD report also mentioned that maintenance problems occur with imported NCMTs due to '. . . reduced interest of the large producers in smaller markets' (TT/67, p. 47). This point can be generalized to other countries with small markets (Jacobsson, 1986, ch. 6).

This last point may suggest that the problem of maintenance does not necessarily solve itself naturally with the passage of time. To the extent that there are significant fixed costs involved in the supply of repair and maintenance services, it may well be the case that foreign suppliers do not care to take these costs for very small markets, Let us therefore discuss further the role of a domestic industry supplying NCMTs in developing countries.

The role of the supplying industry was underlined in the second approach listed in section 2.2.1 above in that it is the supplying industry which develops and produces the new techniques and facilitates the upward movement of the technique on its S-curve. For NCMT, the international machine tool industry has already played this role and the present diffusion of NCMTs in the NICs is in terms of mature techniques. The role of the international supplying industry is therefore not so great for these countries. However, all the NICs concerned, with the exception of Mexico, have an industry that supplies NCMTs. Let us briefly discuss their role in the local diffusion of NCMTs.

With the probable exception of Singapore, the local industry is protected, the import restrictions being either tariffs or quantitative ones. The bulk of the international industry producing NCMTs operates under conditions where economies of scale are important in determining the production costs (Jacobsson, 1986, for CNC lathes; Edquist and Jacobsson, 1985a, for machining centres and Boston Consulting Group, 1985, for CNC lathes and machining centres). The minimum efficient scale of production normally exceeds the total annual market of the NICs as well as the annual output of the NCMT-producing firms in the NICs. In this situation we would expect the prices of the locally made machines to be higher than those available on the international market, although the price difference varies between products and countries. As is mentioned below, there is also some evidence of this in some of the NICs. Obviously, if the local industry is protected and the users have to be satisfied with higher priced goods, the diffusion of NCMTs will be retarded.[5]

On the other hand, a local industry may have very positive effects on the local users. In the literature there is a consensus that an advanced and efficient machine tool industry will have positive effects on the local engineering industry (Jones, 1983; Sciberras and Payne, 1985). Of course, the opposite is the case with an inefficient machine tool industry if it is protected. However, a domestic cost-inefficient industry may also have some positive effects on the local user industry in that the local producer of NCMTs can be expected to perform better with regard to the supply of knowledge and service. UNCTAD TT/67 makes the following statement:

> Domestic producers can be treated as educators in spreading the AMBTs [automatic machine building technologies] in the Yugoslav economy, particularly the biggest producer which organizes regular two week seminars for engineers and technicians, and a few-days seminars for qualified workers, in order to instruct them how to use their equipment and standard software packages. All three firms have permanent maintenance groups. . . (p. 20.)

In the Argentinian case, the local supplier is also pointed out as a provider of training and maintenance for the user (UNCTAD/TT 66). It needs to be stressed, however, that it is not necessary to produce the NCMTs to provide the services. Thus a local industry should not be fostered on account of an expected better performance in this respect than (unsubsidized) foreign suppliers. If the objective is to solve the service/knowledge problem it is this *function* which should be subsidized, whether it is supplied by local or foreign firms or local distributors of foreign machines. To the extent that a local industry is fostered, care should be taken to insure that this is done without increasing the price of the machines and without reducing the access to different types of NCMTs.

9.2.3.5 Interest From the foregoing we can conclude that problems regarding information about, access to and knowledge of how to use NCMTs are clearly retarding the diffusion of NCMTs in the NICs, in particular for smaller and medium-sized firms. In this section we will now discuss, in some detail, the interest condition. We will begin by briefly addressing whether or not the choice of NCMTs may be dependent on strictly technical considerations. Subsequent to that discussion, we will analyse a number of other determinants of the interest of firms in adopting new techniques. The interest condition is closely related to the profit motive of firms in capitalist countries. The profitability of a new technique is, in its turn, influenced by a number of more specific factors, as outlined in section 2.2.1 and the discussion of the three approaches on pp. 134–5 above.

In the literature (e.g. UNCTAD TT/66 and UNCTAD TT/67) the opinion is advanced that purely technical reasons can partly explain the adoption of NCMTs in the NICs. It is even suggested (UNCTAD TT/66, p. 65) that 'in the case of large users fabricating complex equipment, NCMTs are acquired for technological reasons in so far as the product to be manufactured made the use of this technology almost imperative'. Similarly, 'The general conclusion has been derived, that the level of processing sophistication is closely related to the intensity of NCMTs employment. A forklift truck and engine producing enterprise argues that a good number of more sophisticated products could not normally be produced without using the NCMTs' (UNCTAD TT/67, p. 35).

Whilst it may well be so that for a very limited number of products, e.g. aeroplanes, the use of NCMTs may be an imperative, one should be extremely careful in interpreting answers from entrepreneurs as regards their reasons for choosing a particular technique. In very sharp contrast to the Yugoslavian forklift truck and engine producer mentioned in the quotation from UNCTAD (TT/67, p. 35) above, the production engineers of Volvo Components, the Swedish diesel engine producer discussed in section 7.2.2, strongly argue that their products can very well be produced with conventional machine tools and indeed they have a fair number of such machines operating alongside of very advanced FMSs. These production engineers argue that if the operators handling the conventional machine tools are not very skilled, then the precision and regularity of precision for conventional machine tools will be lower than for NCMTs. This however assumes that all the other specific skills needed to operate the NCMTs, e.g. tool choice skills, maintenance skills etc., are available. Thus, an answer in an interview from an entrepreneur in any country has to be considered alongside the absolute level of skills among the workforce that that entrepreneur has at his or her disposal. Indeed, in the Yugoslav case the point is also made that products of uneven quality are produced due partly to reasons of labour skills, which would tend to strengthen this point (UNCTAD TT/67, p. 35).

Of course, provided that a firm has a good quality control department, the problem of uneven quality can be reduced to nil but at a price of high rejection rates. Such rejection rates mean that delivery times may not be met and costs will be high. Differences between techniques in delivery times and production costs obviously influence the profitability of the techniques and are therefore relevant in a discussion of the interest condition.

Given that there is a true choice between conventional machine tools and NCMTs, what then determines the outcome? A large number of factors need to be considered. It is, for example, common that prestige influences the choice of technique, although we would not expect a large scale diffusion of a new technique to be based on such a factor. Increased management control of the production process is another factor of a political nature which can have an economic base. It is not uncommon in India, for example, that better management control is sought – through the introduction of NCMTs and other automatic equipment – in order to increase the utilization of machines (Edquist and Jacobsson, 1985a). Abstracting from such variables for the moment and looking at the choice of technique in a much more conventional and narrow way, we may identify two main factors determining the choice of technique. These are (Jacobsson, 1986, pp. 16–20):

1 The cost of preparing the machine tool (setting it, programming it etc). Let us call this the fixed cost.
2 The cost of cutting the metal, i.e. the cost of actual machining. Let us call this the variable cost. This cost is a function of three factors:
   (a) the cycle time, which is the time it takes for the workpiece to be machined;
   (b) the cost of capital per unit of time, i.e. both cost of the machine and interest rate; and
   (c) the cost of labour for operating the machine tool per unit of time.

In this narrow perspective the choice of machine tool becomes a function of two factors only, namely the cost of preparation (1) and the cost of actual machining (2), i.e. of fixed cost (1) and variable cost (2).

Let us illustrate the choice of technique with the case of lathes. In figure 9.1 we have made an illustrative example indicating the choice of technique where the alternatives are engine lathes, CNC lathes and automatic lathes. The engine lathe is a simple lathe which is fully manually operated. It is very cheap in comparison with other lathes but the cycle time is long, which implies high variable costs, especially when labour is dear. The fixed costs, however, are very low. Given these characteristics, engine lathes have traditionally been used in the production of small batches. The CNC lathe has a shorter cycle time for the same machining task, which may mean that the variable costs are lower than in the case of the engine lathe, especially when labour is dear. The fixed costs have traditionally been higher than for engine lathes due to the need to make computer programs, but the program costs have been reduced substantially in the last decade. In addition, once the computer program has been made, the fixed costs for preparing another batch of the same part after a time interval is close to nil. This means that the more often the production of a particular part or component is repeated, the lower are the fixed costs of using NCMTs in relation to other types of lathe. Indeed, for products that are produced in small batches but with great frequency, e.g pumps, the fixed costs of NCMTs are close to nil and probably below those of engine lathes. The automatic lathe is characterized by very short cycle time for the same machining task and the use of very little labour, implying low variable costs. The fixed costs, however, are very high.

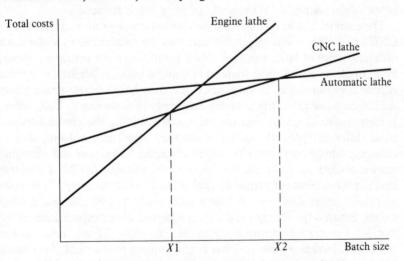

Figure 9.1 Illustrative example of choice of technique between engine lathes, CNC lathes and automatic lathes.

The characteristics of the different types of lathes in our example mean that engine lathes would normally be used for batch sizes smaller than $X1$. CNC lathes would be used for batch sizes in between $X1$ and $X2$. Automatic lathes would be used for batch sizes larger than $X2$. In some situations, however, characterized by a frequent, repetitive production of a part or component, CNC lathes would have lower fixed *and* variable costs than engine lathes.

The precise value of $X1$ and $X2$ would of course vary from country to country and maybe from firm to firm depending on a number of factors. Of immediate concern is the price of labour and the price of capital (machines and interest rate). A relatively high labour price would mean that the variable cost of using engine lathes would rise *vis à vis* the variable costs of the other types of lathe. All the lines in figure 9.1 would then become steeper, but the engine lathe line would be most affected. This would mean that the breakeven point between engine lathes and CNC lathes would be to the left of $X1$.[6] Correspondingly, a relatively low price of labour would mean that the breakeven point would fall to the right of $X1$. On the other hand, a relatively low price of capital would lower the variable cost of using CNC lathes and thus make the breakeven point fall to the left of $X1$. Correspondingly, a relatively high price of capital would push up the variable costs of CNC lathes more than those of engine lathes. Thus the breakeven point between engine lathes and CNC lathes would be shifted further to the right of $X1$. Thus, if we have a situation where (a) the price of labour is relatively low and (b) the price of the CNC lathe is relatively high (for example due to protective regimes of the local machine tool industry), the potential for using NCMTs would be very much reduced.

The exercise can be repeated for the choice between automatic lathes and CNC lathes. We would emphasize here that the relative prices of the three different types of lathe can easily vary greatly between countries owing, for example, to government trade and industrial policies. Whilst a protective regime may not cause the local price of engine lathes to rise much above the international price, this is nearly necessarily so in the case of CNC lathes. It should also be added that the factors determining the choice between these different types of machine tools vary over time, not only due to changing labour costs but also due to changing techniques and changing market conditions. In Japan, for example, the diffusion of CNC lathes was made at the expense of primarily engine lathes until the early 1980s when automatic lathes also began to lose market shares to CNC lathes. In other words, initially, the changes in the technique and labour/capital costs meant that $X1$ was moved continuously to the left whilst $X2$ was more or less stationary. In recent years, $X2$ has began to move to the right (Jacobsson, 1986, p. 212).

The determinants of the real choice of technique, however, are much more complex than is indicated by the narrow discussion presented above, although it does contain important determinants of choice. A further very important determinant, both of the choice of technique and of the efficiency of past choices, is the capacity utilization of the equipment. Let us discuss two determinants of the rate of capacity utilization, i.e. the skill level in the factory and factors of a political and economic nature.

The skill level in the factory determines the capacity utilization in that to be properly utilized NCMTs require some very specific skills. Above we mentioned the need for maintenance and we saw how poor maintenance badly affected the capacity utilization in the Yugoslav case. This argument applies also to skills needed to plan the flow of material, for tool choice and to do the programming. Given that the NCMT involves a considerable investment, factors affecting the utilization of the machine become of critical interest when the costs and benefits of the new installation are calculated. One case in point is described in UNCTAD TT/67:

> A medium sized firm, specialized in the custom designed machinery for food processing, employing 4 NCMTs...about 14 hours per day, has decided to stop purchasing additional NCMTs until it would reach the effective utilization of the present NCMTs at least 18 hours per day in order to reduce the fixed costs of this equipment considerably [*sic* throughout]. (p. 32.)

Factors belonging to the realm of political economy also affect the capacity utilization. Let us take the example of India (Edquist and Jacobsson, 1985a, p. 649). In a number of firms interviewed, management suggested that the efficiency of operators of conventional machine tools was low, resulting not only in a low output per worker but also in a low output per unit of capital. With automated techniques, the control exercised by the operator can be reduced and the utilization of capital increased. This is of particular importance in countries like India where the price of capital goods can well be twice that in the international market. Let us illustrate the effects by one example (see table 9.5). Here we see that the annual cost of using a machining centre amounts to 1.355 million rupees. If we compare with situation (A), i.e. one machining centre replaces three conventional machine tools, the use of a machining centre would not be justified on narrow economic grounds. However, if we assume that, for a number of social reasons, the efficiency in the operation of the conventional machine tools is lower than in situation (A), and that we need to add another conventional machine tool and another labourer per shift, the situation changes greatly. Although the cost per unit of output using a machining centre is still higher than for using conventional machine tools, the difference is marginal. Of interest here is that out of the cost increase between situation (A) and (B), two-thirds is due to increased capital costs. If one machining centre replaces five conventional machine tools, the choice of the machining centre is justified on narrow economic grounds, in spite of the low relative cost of labour in India. Rattner also makes the point, in an example from Brazil (1984, p. 21), that the conventional machine tools used had a much lower capacity utilization than NCMTs.[7]

Table 9.5 Illustrative example of choice of technique in India (million rupees)[a]

	Machining centre	Conventional machine tools		
		(A)[b]	(B)[b]	(C)[b]
Capital costs	1.280	0.640	0.851	1.066
Increased scrap	—	0.086	0.115	0.144
Labour costs	0.075	0.225	0.300	0.375
Total costs	1.355	0.951	1.266	1.585

[a] We have assumed a depreciation time of six years; 18 per cent interest rate and three shift production.
[b] In (A) we assume that three conventional machine tools, manned by one operator each, are substituted for by one machining centre. In (B) and (C) we assume that the number of conventional machine tools substituted for is four and five respectively.
Source: Interview with an Indian machinery producer.

Another factor affecting the choice of technique is the cost of work in progress and stocks that are associated with the various alternative techniques. As Jacobsson (1986, ch. 2) has shown, the inclusion of the variable can significantly affect the choice of technique, although it does not do so in all cases (see also chapters 3 and 5 above).

To summarize this discussion of the determinants of choice of technique we have, in this section, so far pointed to the following factors:

- The batch size normally produced and the frequency with which it is produced.
- Price of labour.
- Price of capital.
- Factors determining capacity utilization (skills and control questions).
- The costs for work in progress, which depends on both the interest rate and the scope for reducing the physical work in progress through using new techniques.

The choice of technique is thus a function of both the type of markets the firms serve, which determine the batch size and the frequency of production, and of reasons relating to factor prices and control factors. As far as the markets that firms serve are concerned, Boon makes the following observation:

> The industrialized countries will have a demand structure which favours complex and specialized tasks, and therefore a heterogeneous, small to medium scale product mix for which the flexible cnc machine is particularly appropriate. That machine therefore has a high and rapid diffusion potential in the industrialized countries. (1985, p. 40.)

On the other hand, Boon suggests that the NICs produce, on average, less complex products and components which are more standardized and mature. This implies that the diffusion potential of the NCMTs would be less in the NICs than in the developed countries (1985, p. 104).

Boon is right to the extent that he makes the very important point that a slower diffusion of NCMTs in the NICs should not necessarily be seen as a lagging behind the developed countries, resulting in a relative technological backwardness (1985, p. 42). He further argues, however, that 'There is no reason why advanced and leading firms in the LDC's and NIC's producing output specifications in batch sizes equal or similar to the ones in the firms of the advanced countries should not use the same methods of production' (p. 42).

The questions raised by Boon are extremely important in assessing the effects of the apparently uneven diffusion of NCMTs globally. The question of output mix and associated batch size cannot be ignored. On the other hand, there is really no empirical evidence showing that the NICs have a radically different output mix than the developed countries. Indeed, most of the NICs are import substituters in the engineering industry and cover very large parts of the domestic demand for capital goods. This would seem to us to suggest that the output specifications are probably not very much different from those in the developed countries. Furthermore, given the inward looking character of the engineering sector in most NICs, and the small size of the domestic market, we would expect that for some industries there would be an extra premium on flexibility in precisely these economies, as opposed to the developed countries, which should lead to a more frequent choice of NCMTs instead of specialized machines, such as automatic lathes. There is some evidence for this. For example, UNCTAD says in the case of Argentina

> The two largest firms claim to have developed more experience in the application of NC equipment for some machining operations than their licensor or parent company. This is explained by the fact that the foreign suppliers of the technology have in their factories at home

more specialized equipment than the firms surveyed in Argentina have. For example, in the case of the machining operation for a key part in a turbine bucket, the Argentine firm concerned has developed special programmes for its NC equipment while the same part is made in the factory of the licensor with specialized but conventional machine tools. (TT/66, p. 25.)

Thus, in this particular case it may well have been that parent company produced much greater volumes than the Argentinian firm and could therefore be justified in using special purpose machine tools, which are relatively inflexible. In terms of figure 9.1, the US firm would lie to the right of $X2$ whilst the Argentinian firm would lie in between $X1$ and $X2$. Similarly, in an interview an engine producer in Taiwan claimed that they bought an FMS instead of conventional transfer line for the same reason.

Although we would not attempt to suggest that the output mix always favours the more intensive adoption of NCMTs in the NICs as opposed to the developed countries, it may well be so for some products which are produced in very large series in the developed countries, but which are produced in small batches in the NICs on account of the orientation of the firms to the requirements of the small local market.

The question of the output mix and its importance for assessing the impact of the new techniques on, for example, international trade, is an empirical one but one that cannot be resolved within the context of this study. This means that the interpretation of the present unequal diffusion of NCMTs must be qualified by this unsolved issue.

We would also argue, however, that relative prices of labour and capital strongly influence the choice of technique for firms producing small and medium-size batches. Evidence from Mexico (Mercado, 1984, pp. 24–5) and Argentina (UNCTAD TT/66, p. 29) shows that although labour is much cheaper in these economies than in the developed countries, labour-saving *is* an important factor explaining the choice of technique. UNCTAD (TT/66) gives the, perhaps extreme, example of an effective reduction in actual machining time from 90 to 11 minutes – from introducing NCMTs – for a batch size of 100–150 units. This means that the cost of labour for machining per unit of output is reduced by 79/90 or around seven-eighths. Even with relatively low wages, such a reduction can have a major impact on the choice of technique. It may also be mentioned that, in our opinion, there is a good possibility that NCMTs would be capital-saving in this particular application. One NCMT replaces seven to eight conventional machine tools in this case, which is unusually high.

The price of machines influences also the choice of technique. In many NICs the price of NCMTs is significantly higher than the world market price. Rattner (1984, p. 19) points to the very high price of NCMTs in

Brazil, as does UNCTAD (TT/67, p. 52) in the case of Yugoslavia. We also referred above to the case of India. Obviously, the relatively high price of NCMTs in relation to labour and conventional machine tools reduces the scope for the profitable use of NCMTs.

9.2.3.6 Concluding remarks The present level and rate of diffusion of NCMTs in the NICs is certainly lower than in the developed countries. However, the potential for diffusion may also be lower in the NICs than in the advanced countries. One possible reason for this is the type of products produced in the NICs. *If* products in the NICs are more simple, more standardized and more mature than those produced in the OECD countries, the potential for diffusion of NCMTs is smaller in the NICs. Another factor that reduces the potential level of diffusion of NCMTs in developing countries is the relative factor prices.

However, the actual diffusion of NCMTs in the NICs is less than we would expect even from this, reduced, potential. We have addressed several reasons for this. The nature of the protective policies for the machine tool industries in the NICs restricts access to, and therefore the scope for the introduction of, NCMTs. Problems regarding information about NCMTs and knowledge about how to use and maintain them further retard the diffusion in some NICs. This is a particularly large obstacle for the smaller and medium-sized firms. In the long term we would expect those problems to be mitigated. However, the factor price relations and possibly also the product mix factor can be expected to prevail in the long run. Therefore, we judge that the potential for applying NCMTs may continue to be smaller in the NICs than in the developed countries.

The main qualification to this argument would be if an NIC faced a greater shortage of traditional skilled machine tool operators than a developed country. As was shown in chapter 3, NCMTs can be highly skill-saving in terms of the skill input of machine tool operators. Of course, other, specialized, skills need to be available if NCMTs are to be properly used, but the 'mass' of skills required to produce a given output may be considerably reduced with the use of NCMTs. To the extent that the NICs face a greater shortage of skilled machine tool operators than do the developed countries, NCMTs could remove an important obstacle for an expansion of production of high quality engineering goods in the NICs. The potential for introducing NCMTs could then be even higher than in the developed countries.

Notes

1 As a percentage of total yearly investment in machinery and equipment in the engineering industry, the figure for Korea in 1982 was 1.25 (table 9.2 and

Economic Planning Board, 1984, table 1.10). For 1983, the figure rose to an estimated 3.28 per cent (table 9.2 and Economic Planning Board, 1985, table 1.14). We have here assumed that the ratio of investment in machinery and equipment to total investment was the same as in 1982, namely 64.1 per cent. Although the equivalent figures for 1984 are not available, the doubling of investment in NCMTs over the 1983 value would suggest to us that the figure continued to rise in 1984. In the case of Yugoslavia the investment in machinery and equipment amounted to US$826 million in 1979 (United Nations, 1980). With an assumed unit price of US$175,000 for NCMTs bought, the share of NCMTs in total investment would be 2.5 per cent. As was shown in table 7.1 column (6), the equivalent figures for the UK, Japan and Sweden were 7–14 per cent in 1983 and over 5 per cent in 1979.

2   We have used the 1980 exchange rate of 7.9 rupees/dollar.

3   The output per employee in the US engineering industry in 1979 was around US$75,000 (United Nations 1980). The output of firms with 100 and 500 employees in the US would then be US$7.5 million and US$37.5 million respectively.

4   We use the term 'structurally weakest' since ISIC 382 is the true machine building industry, demanding a very high technological capability. In the case of Singapore it may, however, be noted that the share of ISIC 382 was only 17 per cent in 1978 and that it is therefore growing very quickly in importance.

5   Above we also discussed how a protected industry may retard diffusion through not allowing local firms access to some versions of the technique.

6   A counteracting factor is that a rise in the cost of labour would increase the fixed costs too, which means that all curves would shift upwards but stay parallel. This upward shift would be largest for automatic lathes and smaller for engine lathes, with CNC lathes normally falling in between. If fixed costs as a proportion of total costs are very large, or if the cost for the labour required to program and set the NCMT is drastically increased, this would possibly result in a move of $X1$ to the right. However, we disregard this possibility in the further discussion.

7   Of course, management may consider the achievement of greater control over the labour process as a sufficient objective for installing new techniques.

——— **Statistical Appendix to Chapter 9** ———
follows overleaf

Appendix table 9.1 Stock of NCMTs in some NICs and their growth rate over time

	1978	1979	(%)	1980	(%)	1981	(%)	1982	(%)	1983	(%)	1984	(%)	1985	(%)
Argentina[a]	116	173	49	251	45	288	15	331	15	380	15	436	15	500	15
Brazil[b]	439	546	24	649	19	773	19	923	19	1,133	23	1,392	23	1,711	23
India[c]	n.a.	n.a.	n.a.	220	n.a.	308	40	430	40	601	40	840	40	1,178	40
Korea[d]	801	942	18	1,017	8	1,110	9	1,288	16	1,712	33	2,145	25	2,738	28
Yugoslavia[e]	656	f								1,232		n.a.		n.a.	

[a] These figures are derived in the following way. For the stock in 1978, we have added the flow data in Chudnovsky, 1984, table 3.1, for the years 1970–5 to the flow data in Jacobsson, 1981, p. 16, for the years 1976–8. Jacobsson, 1981, p. 16, is also used for the years 1979 and 1980, where we have added the flow for these years to the previous years' stock. For the period 1981–5, we have simply assumed that there was an even annual growth rate of the stock of 14.8 per cent which, cumulatively, gives the 1985 stock figures of 500 as reported by Chudnovsky (1986).

[b] The 1978 stock figure refers to the period 1971–8. The data for the period until 1982 were derived by adding the production and import data from tables 1 and 3 in Rattner, 1984, to the previous years' stock/figure. The flow figure for 1982 is estimated in Rattner, 1984. For the period 1982–5, we assume an even annual growth rate of 22.8 per cent p.a. which, cumulatively, gives the 1985 figure of 1,711, as given by Chudnovsky (1986).

[c] For the period 1980–5, we have assumed a cumulative growth rate of 39.8 per cent p.a., which gives the stock figure of 1,178 in 1985 as reported in Central Machine Tools Institute (CMTI), 1986. The figure of 220 in 1980 is also given in CMTI, 1986.

[d] The 1985 figure is slightly higher than the figure in table 9.1 since we have not here adjusted for lack of import data on machining centres and NC metal-forming machine tools, nor have we excluded the 258 cheap NCMTs imported in 1976. See n. (a) to table 9.1. For sources, see table 9.1.

[e] See table 9.1 for source.

[f] For the period 1978–83 the average cumulative growth rate is 13 per cent.

Appendix table 9.2   The development of stock of NCMTs in the
UK and USA, 1971–84

	UK units	UK % growth	USA units	USA % growth
1971	4,731		n.a.	
1972	n.a.	average	n.a.	
1973	n.a.	cumulative 15.5	29,369	
1974	n.a.	growth	n.a.	average
1975	n.a.		n.a.	cumulative 11.3
1976	9,725		n.a.	growth
1977	n.a.		n.a.	
1978	n.a.	average	50,100	
1979	n.a.	cumulative 17.7	n.a.	
1980	n.a.	growth	n.a.	average
1981	n.a.		n.a.	cumulative 15.6
1982	25,802		n.a.	growth
1983	28,669	11.1	103,308	
1984	32,566	13.6	n.a.	

Sources: UK, 1971 – *Metalworking production*, 1977; 1976 and 1982 – Machine Tools Trades
Association and *Metalworking Production*, 1979, 1985, respectively. The stock figures for 1983 and
1984 are derived by adding the flow of these years to the stock of 1982. The flow was calculated
from Machine Tools Trades Association and *Metalworking Production*, 1985. USA, 1973 –
Metalworking Production, 1977; 1978 – Watanabe, 1983. 1983 – Edquist and Jacobsson, 1984.

Table 9.3   Share of value added of the engineering sector in
the manufacturing industry of some NICs, the UK and Sweden

		(1) Manufacturing value added (US$bn)	(2) Engineering sector value added (US$bn)	(3) (2)/(1) × 100 (%)
Argentina	(1983)	n.a.	n.a.	22[a]
Brazil	(1979)	42[b]	15[c]	36
India	(1978)	12[b]	3	25
Korea	(1981)	22	5	23
Mexico	(1979)	29	6	21
Singapore	(1980)	4	2	50
Sweden	(1980)	30	13	43
UK	(1979)	147	57	39
Yugoslavia	(1980)	18	6	33

[a] Share of the engineering sector in manufacturing gross domestic product.
[b] Industrial value added.
[c] Including metallurgy.
Sources: Argentina, UNCTAD TT/66; Brazil, Annuario estadistico do Brasil, 1982; India,
Association of Indian Engineering Industries, 1983; Korea, Economic Planning Board, 1982;
Singapore, Sweden, UK and Yugoslavia, United Nations, 1980.

10

Industrial Robots and Flexible Manufacturing Systems

10.1 Industrial Robots

10.1.1 THE PRESENT DIFFUSION OF ROBOTS IN THE NICs

With the notable exception of Singapore, the diffusion of robots in the newly industrializing countries is still very limited. The available data regarding the stock of robots in some of the NICs are given in table 10.1. In addition to these countries, there are robots in Argentina and Brazil and probably also in India (Chetty, 1982), but the level of diffusion is not well known. According to Tauile (1986), there were 30–40 robots in Brazil in 1985, and a few in Argentina in 1983 (UNCTAD TT/66).

In relation to the total number of employees in the engineering sector, the density of robots in Singapore far exceeds that in Korea and Yugoslavia. It is also noteworthy that the density in Singapore is on a par with that of many of the advanced OECD countries (see table 4.1).

Table 10.1 The stock and density of robots in some NICs[a]

		No. robots	Density[a]
Korea	(1985)	55	1.1
Singapore	(1985)	313	20.2
Yugoslavia	(1983)	32	0.4

[a] Number of robots per 10,000 engineering employees.
Sources: Korea, Edquist and Jacobsson, 1985a; *Industrial Robot*, 1985b. Singapore, Pang, 1986. Yugoslavia, UNCTAD TT/67. The employment figures required to calculate the density are taken from ILO, 1982.

In Singapore, automatic devices of a mechanical and pneumatic nature were used already in the late 1960s (Yang, 1984a). Industrial robots proper were not introduced until 1980 (Pang, 1986). The robot population subsequently increased at a very fast rate; from seven units in 1981 to 313 units in 1985 (see appendix table 10.1). Of the 313 robots, 172 are in the manufacturing industry and 114 are in education services (see appendix table 10.2). As Pang notes (1986, p. 4), the largest single application area of robots is in education, which has 36 per cent of the stock of robots. This clearly indicates the importance given to this technology by the government.

Within the manufacturing industry, it is the electronics industry which has the largest number of robots; indeed, it accounts for 74 per cent of the manufacturing industry's stock of robots. It is noteworthy that most of these are assembly robots (see appendix table 10.2).

All in all, there are 62 organizations which have installed robots in Singapore. Of these, 41 are in the manufacturing industry (appendix table 10.2). Pang informs us (1986) that 23 foreign companies have installed 132 robots leaving 181 robots for 39 Singaporean organizations. Assuming that all the 21 organizations, holding 141 robots, in the non-manufacturing sector are Singaporean, it would mean that 18 local firms in the manufacturing sector have installed 40 robots (appendix table 10.2). Thus, in the manufacturing sector, it is the foreign firms which have installed the bulk of the robots.

In Yugoslavia, of the 32 robots installed, half were of domestic origin. Regarding the industrial distribution of these robots, UNCTAD explains:

> Two initial series of 5 surface painting robots, and 5 welding robots were produced and installed in a home appliance enterprise...and in an electrical machinery enterprise. 7 manipulating robots were produced for a car producing enterprise....Besides, 4 prototypes of robots have been developed by two firms and two research institutes. About 10 robots have been imported and installed in different firms (e.g. foundry, tool production, machine tool parts production etc.). (TT/67, p. 16.)

In Korea, the stock of robots appears to be concentrated in five *chaebuls* (conglomerates). Research institutes such as the Korean Advanced Institute of Science and Technology (KAIST) and the Korean Institute of Machinery and Metals (KIMM) also have some robots. The five conglomerates are Gold Star, which has in the order of ten robots; Daewoo, which also has in the order of ten robots out of which eight are produced in-house; Kia Motors, which has something around nine robots; and Samsung, which has some five robots. Samsung also claims to have sold 12 robots (imported from Japan). Hyundai has around 14 robots. The odd robot is also installed in other firms. All in all, we estimate that the stock of robots was in the order of 55 units in 1985. This may be an under-estimate as at the end of 1983 we estimated the stock to be 35 units (Edquist and Jacobsson, 1985a, p. 650). The 1985 estimate is based on additional information in the *Industrial Robot* (1985b), which cannot be expected to cover fully the Korean robot stock.

As regards the sectoral distribution of the Korean stock of robots, the bulk of them appear to be installed in firms producing automobiles and consumer goods such as TVs and refrigerators.

An interesting feature of the robot diffusion in the NICs is that a fair share of the robots are made locally. This applies also to Argentina. In the Korean case, the government sponsors two projects for robot development. These are undertaken by Daewoo and Samsung. Apparently, Daewoo has made great progress in the development of robots and participated at the robot exhibition 'Robot 9' in Detroit in 1985. It is claimed that a US firm ordered two robots made by Daewoo in conjunction with the fair (*Business Korea*, 1985b, p. 105). Also in Singapore there are firms beginning to produce robots, partly subsidized by the government (Pang, 1985). One had already begun to sell an educational robot in 1985 (Pang, 1986).

10.1.2 DETERMINANTS OF THE DIFFUSION OF ROBOTS IN THE NICs

As far as the determinants of the diffusion of robots are concerned, we will first discuss the obstacles of lack of information, access and knowledge and then consider to what extent firms have an interest in adopting this technique.

10.1.2.1 Information, Access and Knowledge   An important obstacle to a wide diffusion of robots is the behaviour of the suppliers, which affects both the diffusion of *information* about robots and the existence of *knowledge* as to how to install and operate robots. It may also affect the real *access* of firms to the technique. In the case of Korea, several sources refer to a lack of domestic application engineering capability (knowledge) as a major obstacle to the use of robots. One distributor of a Japanese robot claims that the Japanese producer refuses to send people to Korea to solve application engineering problems as the firm is too busy. In one particular case, a sister company to the robot distributor wanted to integrate an NCMT with a robot but could not do so as they did not have the engineering capability. The producer of the robot refused to help. In reality therefore *access* to the technique is hindered (Edquist and Jacobsson, 1985a, p. 560). The general problem is the same in the case of NCMTs. In fast-growing industries, which are intensive in the use of service and/or application engineering, the suppliers may not be interested in sending scarce service or application engineers to marginal markets. Naturally this opens up the possibility for local manufacturers to compete on the basis of a better supply of these services. As was mentioned above, many of the robots installed in the NICs are in fact produced locally. One of the more systematic efforts to manufacture robots takes place in Korea. According to the *Industrial Robot*

> Korea is not likely to be a fruitful market for robots for another five years or so. However, by that time, the main groups [conglomerates] will have developed a range of prototypes, and so will be in a good position to dominate their home market. (1985b, p. 100.)

In Singapore there is a considerable problem of *information* and *knowledge*, especially among the local firms. These firms have so far not bought so many robots, as was mentioned above. Pang writes in this connection that:

> first, small companies and those family owned businesses have no idea about robotics. They rely on labor-intensive technology that they have been accustomed to for decades. They are ignorant of robotics technology and government policies. They will need considerable outside help to robotize. They are slow in responding to technological advancement. The local medium size companies responded to robotization, but they are not utilizing IR effectively due to lack of applications experience and knowhow. The MNCs [Multinational Corporations] are the ones that are well aware of robotics technology because of their parent company abroad. (1985, p. 20.)

The government puts a great deal of effort into trying to diffuse the robot technique, including among smaller firms. As is discussed further below, they subsidize the investment cost and costs of feasibility studies of robots. This does not, of course, solve any information and knowledge problem. The government does, however, sponsor a demonstration/display room with 15 different types of industrial robots. It also conducts regular seminars and trains a large number of engineers and technicians in robot technology (Pang, 1985). The problem of *access* would appear to be less in that there are 22 suppliers of robots in Singapore (Pang, 1986).

10.1.2.2 Interest   The choice of technique between robots and manual labour is to a very large extent a question of straightforward capital–labour substitution. The price of labour in relation to that of capital is then a central factor determining the *interest* of firms in buying robots. Wage costs in some NICs are listed in table 10.2, along with those of Japan, the country with the highest density of robots (see table 4.2). The wage costs in table 10.2

Table 10.2 Annual wage costs of employees in ISIC 384

	Annual wage costs (US$)
Argentina (1980)	1,493
India (1979)	955
Korea (1981)	4,278
Singapore (1981)	4,815
Japan (1981)	16,238

Source: ILO, 1982, tables 12B and 17B. We have assumed that the average Argentinian employee works 40 hours per week and that his/her annual wage costs are equal to the hourly wage times 40 (hours per week) times 52 (weeks per year). In the case of Singapore, we have multiplied the wage costs/hour (given in ILO, 1982, table 17B) with the average weekly hours worked (in ILO, 1982, table 12B) and multiplied this in turn by 52.

are averages for ISIC sector 384, which is the transport equipment sector and the sector in which the OECD countries adopt robots fastest.

Of course, the incentives in the developing countries to introduce robots are on the whole much less than in Japan. The potential for robots in the developing countries can therefore safely be said to be less than in the OECD countries as long as the present labour cost differential is so large.

However, the figures given above are averages and for some categories of workers wage costs may be substantially higher. Again, let us take the Korean case as an example. A spot welder in Korea was said to cost a firm around 6 million won (US$7,500) per annum in 1983 whilst a robot was said to cost about 45 million won (US$56,000). To this figure we need to add around 70 per cent in peripherals and other investment costs. Total investment would then be some 76 million won (US$95,000). If we assume that the robot works two shifts and that it substitutes for two workers in each shift, the payback period, counting only reduced labour costs as benefit, would be 3.2 years. Although the calculation is not complete, it nevertheless indicates that there may be a potential for robots in Korea in some applications. Indeed Hyundai Motors, which in 1985 had automated 35 per cent of the spot welding, either through robots or through other automatic machines, plans to double the use of robots in the next two years in the spot welding shop so that the level of automatic welding will increase to 70 per cent (*Industrial Robot*, 1985b, p. 98). As a comparison, 99 per cent of the spot welding at the Volvo Automobile Corporation in Sweden is done automatically (Cambert, 1984).

Of course, if the relative cost of manual labour increases, the scope for introduction of robots also increases. In Korea, the distributors of robots suggest that the annual market for robots may vary between 50 and 300 units in the second half of the 1980s.

In the case of Singapore, Pang (1986) claims that in six robot-using companies that had calculated the payback period of robot installations, the average amounted to 5.4 years *per shift* per five-day week. To the extent that the robot is used in two or three shifts, the payback period is substantially reduced. This clearly suggests that even in a low wage economy (as compared to the OECD countries) it may be profitable to invest in robots. In Singapore, however, the government subsidizes the installation of robots.

The government has set up a Robots Leasing and Consultancy Company which provides low cost financing of 3.5–4.5 per cent interest p.a. to investors in robots. This is about half the market rate (Pang, 1985, 1986). In addition, Singaporean firms need only pay 10 per cent of the consultancy fee for a feasibility study on the profitability of investing in robots.

As far as other countries are concerned, a study in Yugoslavia suggests that the potential need for robots exceeds 1,000 units in the 1980s (UNCTAD TT/67, p. 16). This demand would materialize mainly for process robots, such as for welding and painting, but also for handling robots. Given the low labour costs in Yugoslavia and that robots substitute mainly for unskilled or semi-skilled labour, such an estimate should be considered with care. Areas where robots could well be installed, however, are those that are physically unhealthy for humans or where workers are not interested in working shifts (UNCTAD TT/67, p. 16). In Singapore, a government study estimates that the local stock of robots will be between 600 and 1,300 units in 1988 (Pang, 1985). With a stock of 313 in 1985, the lower of these two figures will probably be the more appropriate one.

In Argentina, the payback period for a robot installation, with the factor costs prevailing in 1983, would be something between 10 and 15 years. A few robots have been installed, however, by motor companies, for example (UNCTAD TT/66). The reasons are probably either a thirst for knowledge in this area, i.e. a strategic investment in learning, or a technological imperative in the sense that the design of the motor makes the use of spot welding robots essential. In Korea, Hyundai has also introduced some robots to do welding at inaccessible points on a car (*Industrial Robot*, 1985b, p. 98; UNCTAD TT/66).

10.2 The Diffusion of Flexible Manufacturing Systems

As far as FMSs are concerned, only scattered information exists. According to a study from 1983, there is only one FMS installed in India, namely one for producing mechanical transmission systems and material handling equipment. The installation has been supplied by a Japanese machine tool builder and is installed in Elecons factory in Baroda (Perspective Plan Committee, 1983, p. 92). According to industrial sources, Hindustan Aeronautics has another FMS. Both these systems are said to be fairly small – three to five machine tools in each – and of an experimental nature (interview with a representative of a leading machine tool builder).[1]

Several other firms have lines of NCMTs, one firm even with an automatic carrier, but the transfer of workpieces is not automated. The reason is at least partly connected with a desire to avoid controversies with the unions. One firm going for lines of NCMTs is TELCO – a large truck producer – which a few years ago operated in a seller's market. Now TELCO has to make sales efforts and the buyer dictates the design. This new situation has created a need for flexibility in TELCO's production apparatus and their new models are planned to be produced in new systems consisting of a number of NCMTs in lines though not connected in a FMS style. Thus, the changing market situation, coupled with social factors, conditioned a change in production technique. To the extent that the competitive pressure will increase in other industries in India, the example of TELCO will probably be followed by other firms.

As far as Korea is concerned, although a formal survey was not undertaken, it would seem that the use of various types of FMS is not insignificant. Available evidence suggests that it is mainly machine tool firms that have an interest in FMS. Tong-il, the leading machine tool firm in Korea, installed an FMS in their own plant at the end of 1984 (interview). The system is said to include eight of their own machining centres.

Another leading machine tool firm, Daewoo, was in 1984 planning to install an FMS with seven sets of machining centres controlled by a central control unit and linked with transport means in their own plant (interview). A recent report (*Metalworking, Engineering and Marketing*, 1986, p. 148) claims that Daewoo now has two FMSs in their own plants. These were designed in-house. In addition, Daewoo has an FMC of Japanese origin. Daewoo is also exhibiting a small FMC made by themselves. It consists of two of their own CNC lathes and a robot.

A third machine tool firm, KIA, was in 1984 also planning both to use and to supply an FMS system. This firm is owned by an automobile firm and the interest of the machine tool division is based, to some extent at least, on the realization that FMS is largely applied to automobile component production (interview in 1984).

A few, other than machine tool, firms have also begun to implement FMCs. One case in point is Samsung, which has integrated CNC lathes, a machining centre and a robot for the production of Lockheed aircraft engines. The Hyundai Motor Company is also said to have integrated robots with NCMTs for the production of automobile components.

Hence, it would seem that the diffusion of FMSs in Korea has begun. It is noteworthy that it is the local machine tool firms which are taking the lead in the application of the technique. This may eventually lead to a local supply capability of FMSs which will help the further diffusion of such systems to other sectors of industry in Korea.

Of Singapore, Yang says

> Three flexible manufacturing systems, each with a number of CNC machine tools connected to workpiece conveyors as well as loading and unloading robots were installed in two local companies, specialized in precision engineering in Singapore in 1983. (1984b, p. 85.)

Whilst it is not entirely clear whether these are FMSs or FMCs it still indicates a significant interest in this type of technology in Singapore. Cheung (1985) further adds to this impression by reporting 20 FMMs based on CNC lathes with robot arms attached.

Finally, as far as Brazil is concerned, Tauile (1986) emphasizes that there is no FMS in that country.

Note

1   On the other hand, the Central Machine Tools Institute (1986) claims that there is no FMS in India.

—————— **Statistical Appendix to Chapter 10** ——————
follows overleaf

Appendix table 10.1 Breakdown of industrial robots (IR) by applications in Singapore, 1981–5

Year	Assembly	Spray painting	Material handling	Arc welding	Education	IR vendor	Total	Cumulative total	Percentage increase
1981	–	4	3	–	–	–	7	7	–
1982	5	6	5	2	4	–	22	29	314
1983	25	9	12	9	13	8	76	105	262
1984	20	3	10	7	17	5	62	167	59
1985	45	5	3	–	80	13	146	313	87
Total	95	27	33	18	114[a]	26	313		
%	30	9	11	6	36	8	100		

[a] Of the 114 robots classified under education, 49 (43%) are industrial robots and 65 (57%) are desktop robots.
Source: Pang, 1986.

Appendix table 10.2 Breakdown of industrial robots (IR) by manufacturing and non-manufacturing sectors in Singapore, 1985

Industrial code	Sector	No. of companies	IR application						Total robots	No. of robots per co.
			Assembly	Spray painting	Material handling	Arc welding	Education	IR vendor		
384	Electronic products and components	15	95	3	27	3			128	8.5
357	Plastic products	11		21	1	1			23	2.1
381	Fabricated metal products machinery and equipment	8		1	3	9			13	1.6
383	Electrical machinery, apparatus, appliances and supplies	2			1	1			2	1.0
332	Furniture and fixtures except primarily of metal, stone and plastics	2			1	1			3	1.5
372	Basic metal industries	1							1	1.0
342	Printing, publishing and allied industries	1		1					1	1.0
386	Instrumentation equipment, photographic and optical goods	1		1					1	1.0
	Subtotal manufacturing	41	95	27	33	17			172	4.2
931	Education services	11					114		114	10.4
839	Other business services	9						26	26	2.9
955	Repair of personal and household equipment	1				1			1	1.0
	Subtotal (non-manufacturing)	21				1	114	26	141	6.7
	Grand total	62	95	27	33	18	114	26	313	5.0

Source: Pang, 1986.

11

Computer Aided Design Systems

11.1 The Diffusion of Computer Aided Design

In chapter 6, the rate and pattern of diffusion (CAD) in the OECD countries was described. An important feature of the diffusion process was the recent development of CAD software to be used with a personal computer (PC). It was shown that there has been a virtual explosion in the diffusion of this technique in the OECD countries. Corresponding data on the diffusion process in the developing countries are only partly available, that is, for India, Korea and Singapore.

It is clearly the case that CAD is being diffused to the developing countries. The information available on the stock of CAD seats in a number of NICs is summarized in table 11.1.

Table 11.1 Estimated stock of CAD seats in a number of NICs

Argentina	(1983)	74
Brazil	(1985)	340
India	(1987)	770
Korea	(1986)	743
Singapore	(1985)	290
Yugoslavia	(1983)	100

These figures are estimates of varying quality and the reader is referred to n. 1 to this section for a discussion of the weaknesses of the estimates.[1]
Sources: Argentina, UNCTAD TT/66; Brazil, Tauile, 1986; India, Edquist and Jacobsson, 1985a; Korea, Edquist and Jacobsson, 1985a, and Chander, 1987; Singapore, Grumman Int/NTI CAD/CAM Centre, 1986; Yugoslavia, UNCTAD TT/67.

In terms of the rate of diffusion of CAD, we have detailed data only for Korea.[2] Very few firms used CAD prior to 1980. Since then, however, there has been a very rapid diffusion, in particular since 1982. In August 1984

47 firms were reported as having installed large CAD systems, a figure which grew to 73 in 1986 (see appendix table 11.1). All in all, there were 32 large CAD systems installed by late 1983. This number had increased dramatically to 63 systems in August 1984 (Edquist and Jacobsson, 1985a) and to 166 units in 1986 (Kim, 1986). Since the 29 firms having installed CAD up to late 1983 had altogether 32 systems, most firms at that time had only one CAD installation. As the number of systems is now increasing much faster than the number of firms, some firms have apparently acquired more than one unit. This applies in particular to shipbuilding firms and plant engineering firms (see appendix table 11.2). PC-based CAD units are, however, also being diffused in Korea; 79 firms had bought such units by 1986. Prominent among these are machinery firms and firms making printed circuit boards (Kim, 1986, p. 26).

In the case of India, CAD was first applied in the early 1970s by two electronics firms, Bharat Electronics and Indian Telephone Industries. Until 1984, a very limited number of firms had installed CAD and these were mainly in the electronics and aeronautical areas. The odd mechanical engineering and electrical engineering firm, as well as some research and development institutes, had also installed CAD. In 1983, we estimated that the stock of large CAD systems was approximately 25 units with an estimated 100 workstations (Edquist and Jacobsson, 1985a, n. 9).

A fundamental feature of the diffusion process of CAD in India at that time was that the USA, which dominates the industry supplying large CAD units, imposed an embargo on CAD to India. Indeed, at one time there were 60–70 CAD units cleared for import by the Department of Electronics in India but these did not, on the whole, get an export licence from the USA. With the new government of Rajiv Gandhi in India the embargo is not as tight as previously but no CAD units are cleared by the USA government for the nuclear and defence industries. It also still cumbersome to import CAD units for other applications (Chander, 1987). Hence, Indian potential users of CAD simply did not have access to the technology to the extent they wanted.

In 1984, a new electronics company, OMC Computers Limited, was founded in India. The corporate objective of the company was to take a lead in the Indian CAD market. They began by selling a US dedicated draughting system which they soon modified into a multi-user system. They subsequently designed a smaller and cheaper draughting system which is PC-based. The system is sold for 150,000 rupees which is equivalent to less than US$15,000, including the hardware. It is thus a cheap system. The software is totally developed in-house. In March 1987 they were selling 20 units per month. They have now decided to go upmarket and have started to sell a PC-based design unit where the basic software is licensed but where they develop the application-specific software. In the first two months orders were placed for 45 units. OMC computers has today achieved its corporate objective and dominates the Indian CAD market. It is also assessing the possibility of going for exports.

The development of this firm is interesting from several points of view. First, the firm claims that the US quasi embargo is helping them to sell in India. Secondly, the company illustrates that it is not impossible for an NIC-based firm to break into this market. Thirdly, the existence of the firm appears to be very conducive to a local diffusion of cheap CAD units and it would appear as if the Indian market for CAD is now dominated by PC-based CAD. Indeed, India is in the lead among the NICs when it comes to the number of CAD seats installed, even though the data in table 11.1 are less than solid. Noteworthy also is that until March 1987 the company had sold around 450 CAD seats in India. That is, they had by themselves quadrupled the number of seats in India as compared to the situation in 1983.

In terms of the sectoral pattern of diffusion, information is available mainly from Korea and Singapore (see appendix tables 11.2 and 11.3). There is also some evidence from Argentina. In Korea the main users of CAD (in terms of number of systems) are firms in plant engineering, shipbuilding and in the mechanical and electronic industries. In Singapore, the main users are in architecture, structural design and mapping, educational training and precision manufacturing industries (see appendix table 11.3). In Argentina the first CAD units were installed in 1978 by a consulting engineering firm and by a shipyard. In 1983 it was estimated there were a total of 74 workstations in Argentina: 36 (49 per cent) of these were in the mechanical engineering industry; 30 (41 per cent) in engineering services; four (5 per cent) in a shipyard and three (4 per cent) in cartography (UNCTAD TT/66, ch. 5).

The density of use of CAD systems can be roughly measured by dividing the number of systems installed by millions of employees in the engineering sector, which is normally the leading user of CAD. Table 11.2 shows Singapore and Korea to be clearly ahead in this regard, followed by India. One may argue, however, that this indicator understates the Indian density figure due to the notorious overmanning of Indian industry.

Table 11.2 The density of use of CAD in a number of NICs

(1)	*(2)*	*(3)*
	Estimated number	
Country	*of CAD seats*	*Density*[a]
Brazil	340 (1985)	205
Korea	743 (1986)	1,437
India	770 (1987)	527
Singapore	290 (1985)	1,875
Yugoslavia	100 (1983)	140

[a] Number of CAD seats divided by million employees in the engineering sector. For employment data, see table 12.1.

11.2 Determinants of the Diffusion of
Computer Aided Design

We will first look at factors related to *information* about, *access* to and *knowledge* of CAD system. Subsequently, factors affecting the *interest* of firms in adopting the technique will be discussed.

11.2.1 INFORMATION, ACCESS AND KNOWLEDGE

A necessary condition of choosing CAD is, of course, that firms have information about its existence and about is costs and benefits. In Korea, for example, there is a general lack of information about CAD. There is also a lack of trained people to use CAD. A particularly important consideration is that managers are said to be insufficiently aware of the need for and the cost of software development in order to be able to use CAD.[3]

The behaviour of the supplying industry is an important determinant of the diffusion of CAD in the NICs, not only in that it affects the supply of information about its products but also the existence of knowledge of how to repair and maintain the systems. The supplying country and industry may also affect access of NIC firms to their products, as was clearly shown above in the Indian case. The supplying industry may affect the diffusion in less conspicuous ways too. According to an interview in 1984 with Lee Byang Hai, manager of the computer division of Hyundai in Korea, suppliers are 'very cautious' in Korea. The majority do not have repair and maintenance staff in Korea, nor do they have their own representatives.[4] Furthermore, according to Dr Lee Chong Won at the KAIST CAD/CAM Centre, there is no government intervention to ensure that service follows sales. For example, KAIST has a Textronic graphic terminal which broke down in April 1983 and which in November the same year had still not been repaired.

11.2.2 INTEREST

Once the problems of information, access and knowledge are overcome, factors that affect the interest of firms in buying CAD come into play.[5] In chapter 6 the intrinsic or essentially technical reasons for using CAD were discussed. Whilst the situation in the NICs is not necessarily the same with regard to the reasons why firms choose CAD, it is nevertheless so that some reasons are valid for practically all environments. For a number of products, typically electronics products, CAD is an essential – or even necessary – tool and the leading electronics firms in both India and Korea have acquired CAD. Furthermore, modern chemical plants and other process industries are very advantageously designed with CAD, due to the sheer complexity of the systems involved. Plant engineering is, as was seen above, an important user sector in Korea. This diffusion can well be said to be at least partly technically determined.

The ability to meet short lead time is another important reason for choosing CAD. In Argentina, UNCTAD (TT/66) suggests that the shorter lead times associated with CAD were a very important reason for the introduction of CAD by both a capital good producer (with 31 workstations) and a shipbuilder.

Again in Argentina, in the case of shipbuilding, it is argued that

> Delivery time is a crucial competitive factor in the shipbuilding industry and penalties for a delay are very high. For instance, in building a ship valued US$60 million, the penalty for each month of delay could be US$1 million. An investment of US$500,000 in a CAD equipment, therefore, is fully justified. (UNCTAD TT/66, p. 38.)

We also saw that the shipbuilding industry in Korea, which has been very successful in the international market, accounted for 22 per cent of the CAD units in Korea in 1984. The high level of complexity of modern ships may be another reason for introducing CAD.

An obvious factor hindering the choice of CAD is a heavy reliance on foreign designs rather than the generation of own designs. In licence agreements the detailed designs are often also supplied by the licensor. The dependence on foreign licences is widely spread in all NICs. In the Korean case, plant engineering firms and the shipbuilding industry had, in 1984, close to 50 per cent of the CAD installations. Mechanically based firms had only 19 per cent. As the latter include the automobile industry, which has CAD, the remaining firms in the mechanical industry are clearly slow to adopt it. But as the producers do not have much design activity, they simply do not need CAD systems. This was probably true in 1986, when only 16 firms in the mechanical sector (excluding automobiles) had bought large CAD units. They are, however, now buying PC-based CAD to a slightly greater extent. Nineteen firms in the machinery industry have such CAD units according to Kim (1986).

But given that a firm undertakes a lot of basic and detailed design work, under what conditions does it choose CAD? Important considerations here are relative factor prices and skill availability.

As regards the relative factor prices, it is impossible to formulate a general rule of the level of the breakeven point since the productivity increases made possible by CAD vary greatly from situation to situation (see chapter 6). As an illustrative example, we can give Kaplinsky's calculation (1983, p. 110). With a productivity ratio of 3.32:1, and working with double shifts, it was estimated that the breakeven cost of a designer would be US$10,000 per annum. In Argentina the use of CAD was cost-efficient in this narrow sense prior to the massive devaluations in 1981 (UNCTAD TT/66).

Such a narrow cost/benefit exercise assumes, however, that skilled designers and draughtsmen are available for manual design and drawing. In quickly growing NICs which are rapidly changing the structure of their industries, this assumption is not valid. Let us look at the example of Korea and begin with the case of Hyundai, which is one of the largest *chaebuls* and also the largest user of CAD in the country. Between 1975/6 and 1983 they installed altogether nine large CAD units. Their reasons were explained as that the main advantage of CAD was that its software embodied accumulated design and draughting experience. Korea has a shortage of experienced designers and draughtsmen and CAD is seen as a means of catching up with the OECD countries. Furthermore, for Hyundai, the actual cost of this lack of designers lay in the need to use consultancy firms, which charge US$50–60 per hour. For Hyundai, productivity of designers increased by a factor of four to five after six months, which indicates a tremendous saving in manpower. This means there is a need to generate only 20–25 per cent of the number of experienced design engineers per unit of output as compared to non-adoption of CAD (Edquist and Jacobsson, 1985b, p. 560).

This view is partly confirmed in a study made by KAIST in Korea. In a survey of 29 firms using CAD, the most frequently mentioned (pre-investment) motive for buying CAD was to increase the level of technology of designs. The *post*-investment experience was somewhat different. Reduced lead time and the ease of making changes in the designs ranked highest. Thereafter came the more technologically orientated factors – improvement in the technological level of designs and improved standardization of items. Better blueprints ranked next and only thereafter came the traditional cost reducing factor of improved labour productivity (KAIST, 1984). Of course, the ease of making design changes is partly a matter of reducing labour costs for design modifications, but it is also a question of reducing the lead time. Hence, discussion of the interest of firms in using CAD requires consideration not only of relative factor prices but also of the availability of skills at the going rate. Lead time is a further factor that goes beyond relative factor prices as a determinant of choice of technique.

Notes

1   The stock of CAD in the NICs has been estimated in the following way. In the Argentinian case, no formal survey has been made but the data are based on UNCTAD's (TT/66) updating of an older paper on the subject written in Argentina. In the Indian case, Edquist and Jacobsson (1985a) estimated that there were 25 large CADs in India in 1983. As India is still subjected, at least partly,

to a US embargo on super-minis, this figure is assumed to have increased to 30. Each of these is assumed to have four workstations. In addition, an estimated 650 PC-based units existed in India in early 1987 (Chandler, 1987). For Singapore the data come from Grumman Int. NTI CAD/CAM Centre (1986). The data are probably an under-estimate of the true stock since they are based on a survey of only 30 selected users. In 1984, Yang suggested that there were 200 workstations in Singapore (1984a). The data on Korea are based on work by Kim (1986). Each large CAD is assumed to have four workstations. The data are probably an under-estimate of the actual stock of CAD in Korea since we have only partial information on PC-based CAD. According to Kim (1986), there are 79 firms using PC-based CAD but there is no information on the number of such systems. We have assumed that each firm has one system. For Yugoslavia, the source is the following quotation from UNCTAD: 'CAD has recently entered the Yugoslav capital goods sector and other spheres of the economy. Among the users of CAD in the engineering branches (about 25 units) there are the largest and above-average export oriented firms. Up to the present, CAD technology has been used, first of all, for projecting, constructing, forming, and for structural calculations and simulations. However, some of the most advanced users, combining standard software packages with their own software building, are making efforts into the phase of CAD-CAD...' (TT/67, p. 17.) It is not entirely clear if the bulk of the CAD units are in fact interactive, especially those used for calculations. We assume that there are 25 CAD systems and each have four workstations. The Brazilian data are from Tauile, 1986.

2   This section on Korea is based on studies by Edquist and Jacobsson (1985a, p. 650) and Kim (1986).

3   For example, to the extent that imported software presupposes the use of a specific raw material which is not available in the country, the user has to modify the software.

4   It appears as if it was not until 1984 that Computervision, the world leader in sophisticated CAD systems, gave distribution rights to a Korean company, Seoul Electron Co. (*Business Korea*, 1985a, p. 33). Calma, another leading supplier, does, however, sell CAD through the electronics firm Samsung Semiconductor Company.

5   The following section is based on Edquist and Jacobsson, (1985a, p. 651).

Statistical appendix to chapter 11

Appendix table 11.1 Number of firms having installed large CAD systems in Korea

Year	No. systems
Pre-1980	2
1980	5
1981	11
1982	19
1983	29
August 1984	47
1986	73

Source: Edquist and Jacobsson, 1985a, p. 648; Kim, 1986, p. 17.

Appendix table 11.2 Areas of application of large CAD systems in Korea

Area	No. systems[a]	% all systems	No. firms[b]	% all firms
Shipbuilding	14	22	4	5
Mechanical	12	19	18	25
Electronic	11	17	19	26
Construction	4	6	5	7
Plant engineering	15	24	2	3
Others (architectural, educational etc.)	7	11	25	34
Total	63	99	73	100

[a] August 1984.
[b] 1986.
Sources: Edquist and Jacobsson, 1985a, p. 648; Kim, 1986, p. 26.

Appendix table 11.3 Areas of application of CAD in Singapore, 1984

Area	No. stations	% all stations
Architectural, structural design and mapping	108	54
Educational training	46	23
Precision manufacturing	28	14
Electrical and electronics	8	4
Shipbuilding and marine	8	4
Garment manufacturing	2	1
Total	200	100

Source: Yang, 1984a, p. 52.

12
Implications for the Developing Countries

This chapter will discuss the implications for the developing countries of the technological transformation addressed in chapters 3 to 11. Section 12.1 summarizes the data on the diffusion of the various flexible automation techniques. In sections 12.2 and 12.3, implications of this diffusion for the competitive strength of the developing countries in the engineering sector is discussed. General implications are treated in section 12.2, whilst section 12.3 addresses the impact of product level. Finally, some policy implications will be discussed in section 12.4.

12.1 Summary of the Diffusion of Flexible Automation Techniques in OECD and Newly Industrializing Countries

The four flexible automation techniques addressed in this book represent one of the most significant technical transformations of the engineering industry in this century. The four techniques are being diffused at a rapid pace in the developed countries, and while there is also diffusion in the developing countries, it is less pronounced and is restricted to the more advanced developing countries.

Table 12.1 summarizes the available data on the stock and density of the various flexible automation techniques in some OECD economies and certain advanced developing countries.

As far as the *stock* of the flexible automation techniques is concerned, for NCMTs it varied between 6,010 (Sweden) and 118,157 (Japan) among the developed countries in 1984. Among the developing countries, the stock varied between 500 (Argentina) and 2,680 (Korea) in 1985. For robots, the stock in the developed countries amounted to around 100,000 in 1984 and the addition to the stock in that year was 30,000. The stock in the developing

Table 12.1 The stock and density[a] of flexible automation techniques in a number of OECD and developing countries (stock divided by million employees gives density)

		(1) NCMTs	(2) Robots	(3) CAD	(4) FMS
FRG	stock	46,435 (1984)[b]	6,600 (1984)	11,000 (1983)	25 (1984)
	density	11,376 (1980)	1,617 (1980)	2,694 (1980)	6 (1980)
Japan	stock	118,157 (1984)[b]	64,657 (1984)est.	7,300 (1984)[c]	100 (1984)
	density	22,399 (1980)	12,257 (1980)	1,384 (1980)	19 (1980)
Sweden	stock	6,010 (1984)	1,900 (1984)est.	1,900 (1984)[d]	15 (1984)
	density	22,177 (1980)	7,011 (1980)	7,011 (1980)	55 (1980)
UK	stock	32,566 (1984)	2,623 (1984)	9,000 (1983)	10 (1984)
	density	10,505 (1980)	846 (1980)	2,903 (1980)	3 (1980)
USA	stock	103,308 (1983)	13,000 (1984)est.	59,400 (1984)[e]	60 (1984)
	density	11,728 (1980)	1,475 (1980)	6,743 (1980)	7 (1980)
Argentina	stock	500 (1985)	n.a.	74 (1983)	n.a.
	density	n.a.	n.a.	n.a.	n.a.
Brazil	stock	1,711 (1985)	35 (1985)	340 (1985)	0 (1986)
	density	1,033 (1980)	21 (1980)	205 (1980)	

India				
stock	1,178 (1985)	n.a.	770 (1987)	0– 2 (1984)
density	807 (1980)	n.a.	527 (1980)	0– 1 (1980)
Korea				
stock	2,680 (1985)	55 (1985)	743 (1986)	3 (1985)
density	5,176 (1980)	106 (1980)	1,437 (1980)	6 (1980)
Singapore				
stock	700 (1985)	313 (1984)	290 (1985)	0– 3 (1984)
density	4,526 (1979)	2,024 (1979)	1,875 (1979)	0–19 (1979)
Yugoslavia[f]				
stock	1,232 (1983)	32 (1983)	100 (1983)	n.a.
density	1,720 (1979)	45 (1981)	140 (1979)	n.a.

[a] See corresponding note to table 7.2.
[b] See corresponding note to table 7.2.
[c] See corresponding note to table 7.2.
[d] See corresponding note to table 7.2.
[e] See corresponding note to table 7.2.
[f] The number of employees refers to the socialized sector only.

Sources: For the industrialized countries see table 7.2. For stock figures of the techniques in the developing countries: *NCMTs*, table 9.1; *robots*, table 10.1 and section 10.1.1; *CAD*, tables 11.1 and 11.2; *FMS*, section 10.2. For employment figures: industrialized countries, table 4.1; Brazil, Annuario Estadistico do Brasil, 1982; other developing countries, ILO, 1982, table 5B. The employment figures given for the developing countries are: Brazil, 1,657,000 (1980); India, 1,460,000 (1980); Korea, 517,000 (1980); Singapore, 154,660 (1979); Yugoslavia, 716,000 (1979).

countries listed in table 12.1 amounted to around 400 units. For CAD, the stock in five developed countries (USA, FRG, UK, Japan and Sweden) amounted in 1983/4 to around 89,000 seats whilst in the developing countries listed, the stock amounted to around 2,300 seats. Finally, for FMS, the stock in five developed countries (Japan, USA, FRG, Sweden and the UK) amounted to 210 units in 1984 whist there were only a few units in the developing countries.

A comparison in sheer numbers between economies of different sizes has, of course, to be supplemented with data that are normalized. In table 12.1 we have therefore also calculated a density measure for each of the flexible automation techniques. The density measure used is the number of the various flexible automation techniques installed divided by millions of employees in the engineering sector. This denominator was chosen in lieu of value added as it makes the analysis independent of the well-known problems of comparing value added in economies that broadly follow a free trade policy with those that follow a protectionist policy. The use of number of employees as the denominator may, however, well be argued to under-estimate the density in some developing countries, especially India, because of varying degrees of overmanning.

Table 12.1 shows that the difference in the density of the flexible automation techniques is indeed very considerable and that the flexible automation techniques have not been diffused to the same extent in the newly industrializing countries as in the OECD economies. Table 12.2 shows the differences for two groups of countries within table 12.1. The ratios between the densities of these two groups of countries (where the value of the OECD countries accounts for the nominator) were 8.5 for NCMTs, 8.3 for CAD and 43.0 for robots. Although the data base is not without problems, the order of magnitude is clearly there to be seen.

Table 12.2   Density[a] of units of flexible manufacturing techniques in five NICs and five OECD economies

	(1) NICs[b]	(2) OECD countries[c]	(2)/(1)
NCMTs	1,665	14,230	8.5
CAD	498	4,114	8.3
Robots	96	4,122	43.0

[a] Number of units divided by millions of engineering employees.
[b] Brazil, India, Korea, Singapore and Yugoslavia.
[c] FRG, Japan, Sweden, UK and USA.
Source: As for table 12.1.

The data show furthermore the very substantial difference between the various flexible automation techniques with respect to the position of the NICs *vis à vis* the OECD economies. In terms of the diffusion of NCMTs and CAD, the NICs are relatively better off than in terms of robots. This is probably due to the fact that NCMTs are the most mature of the four flexible automation techniques, but the fact that NCMTs are skill-saving is also important. However, the developing countries are not catching up in terms of the growth rate in the stock of NCMTs where they surpass the OECD countries only slightly. Jointly, the developing countries in appendix table 9.1 had a cumulative growth rate of the stock of NCMTs of 23 per cent in 1978–85 whilst that of the USA in the same period was 16 per cent and that of the UK in the period 1976–83 was 18 per cent (see appendix table 9.2). Furthermore, in terms of the share of NCMTs in total investment in machine tools they are *far* behind. Whilst the range in the NICs listed in table 9.2 was 7–23 per cent, the range in the developed countries was between 40 and 62 per cent in 1984 (see table 3.2).

The reasons for the rapid diffusion of CAD are similar to those for NCMTs. For CAD the relative position of the NICs has improved very much during the past few years, i.e. since PC-based CAD software emerged. (The ratio has decreased from about 30 to 8.3 in a few years.)

As would be expected, among the three techniques – NCMTs, CAD and robots – the NICs are worst off in terms of robots, since these save mainly unskilled labour which is abundant in most developing countries. South Korea and Singapore may be two exceptions since unemployment is quite low in these economies. In the case of Singapore, considerable state intervention has also contributed to the fact that robot density is greater than in some OECD countries such as the FRG, the UK and the USA (see table 12.1).

Only a few FMSs have been installed in the NICs. The main reason is probably that the technique is immature and very complex. Even in the leading OECD countries there are many technical problems associated with the installation and running of these systems.

There is also a considerable difference between the NICs as regards the use of flexible automation techniques (see table 12.1). Korea has the largest number of NCMTs, Singapore has the largest number of robots and India has the largest number of CAD seats. Singapore is clearly the NIC with the highest density of robots and FMSs, but Korea holds this position for NCMTs (table 12.1). It is to be noted that concerning robot density Singapore has surpassed the OECD countries with the exception of Japan and Sweden. For FMS, it is noteworthy that Korea has a density which is roughly the same as that for the USA, the UK and the FRG. If we use the higher estimate, Singapore is even on a par with Japan in this respect.

12.2   General Implications for the Developing Countries

Some *general* implications for the developing countries from what has been said hitherto will be discussed in this section. However, it needs to be emphasized that above only a few of the developing countries were addressed. The technological transformation that has been described will surely have very important implications also for the majority of the developing countries that we have considered (Edquist, 1986, pp. 1–13).

A typology of developing countries worked out in 1985 by the United Nations Industrial Development Organization (UNIDO) in relation to its second consultation on capital goods is useful in this context. In this typology 91 developing countries – for which data were available – were divided into three broad groups (A, B and C) utilizing indicators such as the value of capital goods production and imports, manufacturing value added etc. (The typology was presented in UNIDO, 1985b and UNIDO, 1985c.) The three groups are:

Group A   Countries with a very large domestic market for capital goods, very large capacity for capital goods production and high bargaining power (i.e., newly industrializing countries). Seven countries were identified: Argentina, Brazil, China, India, South Korea, Mexico and Singapore. (It may be noted that India and China are often not included in the NIC group, but that Hong Kong and Taiwan often are.) These countries have a fairly well-developed capital goods industry.

Group B   Countries with a medium-sized market for capital goods, medium-to-large capacity for capital goods production and medium bargaining power – 30 countries. These are countries that have started to establish their industrial base with some capital goods industries: examples are Malaysia, Thailand, Colombia and Venezuela.

Group C   Countries with a small-to-medium-sized market for capital goods, very low-to-low capacity for capital goods production and very low-to-low bargaining power – 54 countries. These countries have no or only an embryonic capital goods industry.

Including the OECD countries, we have thus four categories of countries – disregarding the industrialized planned economies in this context.

The engineering sector is undergoing a fundamental technical transformation. As was shown above, this transformation is taking place much more rapidly in the OECD economies than in the NICs and, by assumption, in the countries constituting Groups B and C. Given the benefits that accrue to the investors in flexible automation techniques, this clearly means that the OECD countries have strengthened their position *vis-à-vis* the developing countries as a whole.

One important qualification to this conclusion as regards NCMTs only, would be Boon's argument (1985), presented in chapter 9 on p. 147, that NCMTs are not used to the same extent in the developing as in the developed countries simply because of the nature of the goods produced in the developing countries. Depending on the degree to which this argument is valid, the uneven level and rate of diffusion would therefore not imply a strengthening of the competitive position of the OECD countries.[1]

The impact on competitiveness of the diffusion of flexible automation techniques tends to differ between the four groups of countries.

1 The main *winners* in this process are the OECD countries and Group C countries. The former have increased their competitiveness. The latter, which hardly produce any capital goods, have got access to cheaper capital goods in the short run (through imports) than would have been the case if the technical transformation had not occurred. (However, the possible entry of these countries into capital goods production, in the long run, has become more difficult.)

2 The main *losers* are probably Group B countries. The flexible automation techniques seem not to have been diffused to these countries to any significant extent. This means that these countries will not benefit from the productivity increases associated with the use of flexible automation techniques, in sharp contrast to the OECD countries and the NICs. At the same time, the advantages that these countries may have in terms of cheaper labour are partly eroded by the use of flexible automation techniques in the OECD countries and in the NICs.[2]

3 In between losers and winners are the countries comprising Group A. The advantages that these countries have had in terms of cheaper semi-skilled labour are being eroded, especially due to the diffusion of robots and FMSs in the OECD countries. Furthermore, these countries have adopted the new techniques to a lesser degree than the OECD countries, which means that they are able to draw on this source of productivity increases to a lesser extent than the OECD countries. However, these countries can benefit from the skill-saving character of both NCMTs and CAD and combine cheap engineers/technicians with the new techniques to achieve a competitive edge on the basis of relative factor costs. On the whole, however, these countries will need to follow suit in the technical transformation in order to be competitive with the OECD economies.[3] Furthermore, the production engineering skills associated with system development need to be generated in these countries (see further section 12.4). As was emphasized in chapter 7, however, the relative prices of generally available factors of production comprise only one of several determinants of international competitiveness.

12.3 Specific Implications for the Developing Countries

As was underlined in section 7.1, the magnitude of the impact of the new techniques varies depending on the type of products that are being discussed. A number of product groups were listed which are judged as being most affected by the new techniques. The product groups were: automobiles, heavy electrical equipment, cutting tools, pumps, valves, compressors, construction machinery, tractors, machine tools, special industrial machinery and electronic components. The impact also varies depending on how important production costs and other determinants – e.g. firm-specific assets – are in determining international competitiveness in the various product groups. Typically, the structure of the industries producing these products is characterized by product differentiation and firm-specific assets and less by price competition.

The product groups mentioned are very important in the structure of *production* in some of the NICs. Whilst in Sweden in 1983 these product groups accounted for 68 per cent of the gross output of the engineering sector, in India they accounted for approximately 54 per cent and in Korea for between 54 and 69 per cent.[4] The developing countries' efforts to improve their performance in these product groups are, generally speaking, coming up against a new obstacle, namely the improving competitiveness among the OECD countries due to a rapid diffusion of flexible automation techniques.

For the developing countries, however, the structure of *exports* does not always reflect well the structure of production. On the whole, the simpler type of goods are exported whilst the more complex goods are produced within the framework of an import substituting policy. Let us take the case of Korea as an illustrative example. We made a simple regression at the four digit Korea Standard Industrial Classification (KSIC) level where the dependent variable was the net export ratio (for definition, see below) and the independent variable the intensity in use of engineers and technicians, as measured by the number of engineers and technicians in total employment. The year for which the analysis was made was 1980. The result was that the net export ratio was significant at the 1.2 per cent level and negatively dependent on the engineer and technician intensity. The R^2 was 0.33.

This clearly suggests that the Korean engineering industry's international specialization lies in simpler goods. It does not, however, tell us anything about the character of those goods which are produced under an import substitution scheme.

In order to assess the character of those products that are basically produced in an export-led fashion and those that are produced under an import substitution scheme, we undertook a detailed study of Korea's input/output tables of 1975 and 1980. We classified the sectors, at the lowest statistical level, according to their values for the three indicators of international specialization given below.

1 The *net export ratio* which is measured by $(X-M)/(X+M)$. The ratio can vary between -1 and $+1$.
2 The *export share* of production which is measured as X/P and can vary between 0 and $+1$.
3 The *import share of investment* which is measured by M/AC, where AC is apparent consumption or production minus exports plus imports. The ratio can vary between 0 and $+1$.

The net export ratio (1) measures Korea's specialization at the border, which means that only that part of Korea's output that is subjected to international competition is analysed. The export share (2) is simply the share of output which is sold on the international market. The import share (3) shows the specialization of Korean firms *vis à vis* foreign producers on the Korean market (Ohlson, 1976). We classified the sectors in terms of having a high, medium or low net export ratio, export share and import share. For the net export ratio we defined low as less than -0.25, medium as between -0.25 and $+0.25$ and high as higher than $+0.25$. For the export and import shares, low meant less than 0.25, medium between 0.25 and 0.5 and high larger than 0.5.

On the basis of these three indicators we can identify four different groups within the Korean engineering industry which vary fundamentally in their net export ratios and in their export and import shares. In addition, we have calculated what we call the *marginal export ratio* (MER) for the four groups. This ratio is the share of the increase in production between 1975 and 1980 that was exported. The MER is seen as an indicator of the extent to which the industrialization in this period was of import substituting or export oriented character.

Group 1   In this group, which accounted for 27.4 per cent of the production value in the Korean engineering industry in 1980, we have 11 branches which all are characterized by

- a high net export ratio;
- a medium to high export share;
- a low import share.

Thus, the industries in this group have large trade surpluses, an important orientation to the international market and they compete successfully with foreign products in the local market. The MER, as we would expect, is high

– 48.3 per cent of the growth in production in the period 1975–80 was sold in the international market.

The types of products produced in these industries are basically of three kinds. One subgroup consists of simple metal products. The sectors within this subgroup are household metal products, metal doors, iron wire products, structural metal products, metal cans and shipping containers. The second large subgroup consists of four sectors: electric lamps, radios, TVs and other electronic sound appliances. Finally, in Group 1 we have the sectors covering coils and transformers and wooden and other ships. These products are generally *characterized by a technological simplicity* in terms of their intensity in use of engineers and technicians.

Group 2   There are 15 branches in this group, which accounted for 30.1 per cent of the value of production in the Korean engineering industry in 1980. They are characterized by

- a low export share;
- a low import share;
- a low to high net export ratio.

These products are thus traded to a very little extent. The net export ratio varies from low to high but there is a large element of chance in the determination of the net export ratio given the marginal character of the trade. The net export ratio is therefore of little interest to us in this group. The MER is very low, 12.7 per cent, which clearly suggests an import substituting character of this group.

Within this group we can find, on the one hand, home electrical appliances (refrigerators, laundry appliances, electric fans and other household equipment), and on the other, large parts of the transport industry (railroad vehicle parts, passenger motor cars, buses, trucks, bicycles and parts, motorcycles and parts as well as farm machinery). In this group we also have metal fittings, insulated wires and cables, batteries and other electronic parts and components.

The products in this group are generally of *medium to high technological complexity*. The product groups within the transport industry are those that are greatly affected by flexible automation techniques in OECD countries, the automobile industry in particular.

Group 3   There are 16 branches in this group, which accounted for 22.0 per cent of the output of the Korean engineering industry in 1980. These are characterized by

- a low export share;
- a medium to high import share of investment;

- a strongly negative net export ratio;
- an MER of only 10 per cent.

Thus, an inward orientation of the industry is coupled with a high import ratio. The types of products within this group are mainly mechanically or electrically based machinery. The sectors are: metal-cutting machinery, metal processing machinery, construction and mining machinery, textile machinery, other general industrial machinery, office and service machinery, sewing machines, other industrial machinery, general machinery parts, power generating machinery, dynamos and motors, transformers and other electrical industrial apparatus. Some other sectors are also included in this group, namely motor vehicle parts, railroad vehicles and wire communication equipment.

The products in this group are typically *highly intensive in use of engineers and technicians* and constitute the 'true' machine building industry. It is here we find those products which are judged to be *most* affected by flexible automation in the OECD countries.

Group 4 There are 14 branches in this group, which accounted for 20.5 per cent of the production of the Korean engineering industry in 1980. They are characterized by

- a low to medium net export ratio;
- a medium to high import share of investment;
- a medium to high export share;
- a very high MER, 67.8 per cent.

The dominant category in this group is electronically based products, where we would expect a great deal of re-export of imported parts and components after assembly (electric transmission equipment, wireless communication equipment, electronic calculators, electronic tubes, semiconductor elements, integrated circuits, resistors and condensors). We also find woodworking machinery and steel ships in this group as well as other transport equipment, ship parts, other metal products, bolts, nuts, screws and tools. In terms of technological complexity, this group is somewhere in the middle.

Hence, we have two groups which can be called import substitutors (Group 2 and Group 3), and which accounted for 52.1 per cent of the production in value of the Korean engineering industry in 1980. Two groups are basically export oriented (Groups 1 and 4) and accounted for the remaining 47.9 per cent. The import substituting industries are clearly those of higher technological complexity whilst those goods produced under an export oriented scheme are relatively simple ones. Furthermore, it is mainly among the import substituting industries that we find the greatest effect of flexible automation techniques in the OECD countries.

In the longer run, however, Korean, and other advanced developing countries' ambitions lie in reaching international competitiveness in the technically more complex products also. Given the slow diffusion of the new techniques to the developing countries, we believe that it will be even more difficult for the developing countries to realize this ambition after the current rapid automation of the engineering industry in the OECD countries. This means not only that the developing countries will have additional problems of earning foreign exchange through exporting more complex capital goods. It also means that to the extent that the unequal global diffusion of the new techniques is associated with differential productivity increases, the social cost of fostering the production of more complex capital goods for the home market in the developing countries will increase.

As far as those product groups that are produced under export schemes are concerned, the Korean example suggests that they are on the whole not greatly affected by flexible automation techniques. This would suggest that the flexible automation techniques do not pose a threat to the competitive position of the Korean producers in their present chief lines of export in the engineering industry. Let us broaden the discussion, however, to include the other developing countries and then investigate whether the result from Korea is generally valid.

It is probably true that the export pattern often does not reflect the structure of production in the engineering sector of most developing countries. On the whole, the export share of production, and the import share of investment of engineering goods, is very low in the major capital goods producing NICs. This can be seen in table 12.3. With the exception of Singapore, there is therefore no reason why exports should give a good reflection of the structure of production.

In appendix table 12.1, data are presented for those product groups that were previously judged to be most affected by flexible automation techniques, as regards their significance in all the developing countries' *exports*. Column (1) presents the share of the developing countries in world export in 1983. Column (2) presents the percentage by which the export growth of the developing countries exceeded the growth of exports of the world in 1970–83. On *average*, the developing countries' exports of *all* engineering products grew 107 per cent faster than the world exports in this period (UNCTAD trade data bank). In column (3) the share of the particular product group in the developing countries' total manufacturing exports is presented.

Three observations can be made for the product groups (listed in appendix

Table 12.3 Production and trade in machinery and transport equipment in some developing countries, 1978–80

	Export/ production (%)	Import/ apparent consumption (%)
China	1	5
Brazil	5	12
Mexico	4	29
Korea	23	38
India	5	13
Singapore	91	93

Source: UNCTAD TT/65.

table 12.1) that show the greatest impact of flexible automation techniques in the OECD countries.

1   Generally speaking, *the product groups show a low share for the developing countries in world exports* (column 1). The main exceptions are SITC 7293 (thermionic valves and tubes, transistors etc); SITC 7221 (electric power machinery); SITC 7118 (engines not elsewhere classified), which is very small in terms of value of exports; and SITC 7115 (internal combustion engines, not for aircraft).

2   Generally speaking, *the growth rate in exports of the product groups listed in the table is inferior compared to the average* (107 per cent) for all product groups in the developing countries (column 2). The exceptions are SITC 7341 (aircraft); SITC 7125 (tractors); and SITC 7221 (electric power machinery).

3   With the exception of SITC 7293 (thermionic valves, transistors etc.), *the product groups listed are in the low to medium importance range in terms of their share in the developing countries' manufacturing exports* (column 3).[5]

Thus, *those product groups where flexible automation techniques are having the greatest impact in the OECD countries are, generally speaking, among those where the developing countries have been less successful in terms of exports.* This is probably simply a reflection of the high technological complexity of those products that are mostly affected by flexible automation techniques. In the Korean case, these products were produced within an import substituting scheme and the same probably applies for the NICs in general. The relatively poor export performance is therefore probably due to both the technical complexity of these product groups and to the fact that they are produced within the framework of an inward looking industrialization. We cannot, however, exclude the possibility that the relatively poor export performance can at least partly, and particularly during recent years, be a function of improved competitiveness of the OECD countries due to the adoption of flexible automation techniques.[6]

In contrast, appendix table 12.2 shows those engineering product groups in which the developing countries have captured a significant share of the world export market. These products would appear to be more simple ones, many falling into Groups 1 and 4 in the analysis of the Korean industry above. In these product groups the impact of flexible automation techniques has hitherto not been great.

We can thus draw four conclusions:

1   Those product groups that show the greatest impact of flexible automation techniques in the OECD countries constitute a large share of the value of production in the engineering sectors of the NICs. Therefore, the impact on the NICs is potentially great.

2   It would seem as if the product groups that are mostly affected by flexible automation techniques in the OECD countries are predominantly produced under import substituting schemes in the NICs. The Korean case shows this in a very clear way. The export performance of all developing countries for these products is relatively poor. The weak export performance is probably mainly due to the fact that these product groups are technically complex. We cannot, however, exclude the possibility that flexible automation techniques have already had an impact on the international competitiveness of the OECD countries at the expense of the developing countries.

3   Given the slow diffusion of flexible automation techniques to the NICs, we believe that their ambitions to become internationally competitive in the product groups that are now produced in an import substituting way are running up against a new obstacle – the improving competitiveness of the OECD countries due to a rapid adoption of flexible automation techniques. This means that the developing countries will have greater problems in competing in the international market, but also that the social cost of fostering the production of more complex engineering products in the developing countries will rise.

4   For those product groups in the engineering sector where the developing countries have had a good export performance, the impact of flexible automation techniques is relatively minor. The uneven global diffusion of flexible automation techniques therefore does not constitute a significant threat to the present export performance of the NICs in the engineering sector. On the whole, the product groups where the developing countries have performed well are technologically simple.

12.4   Policy Implications for the Developing Countries

Flexible automation techniques *will* be the dominant techniques in the engineering industries of the developed countries in the future. At the same time, we have seen that the diffusion of flexible automation techniques in the developing countries is very much lagging behind that in the OECD countries. We have also suggested that this unequal diffusion will have negative effects on the developing countries' ability to become competitive in those product groups where the new techniques have the greatest impact. What does this imply for the developing countries with regard to policy and strategy? Should they ignore the new techniques or should they support their diffusion?

The concept of 'social carriers of techniques,' which was presented in section 2.2.3, identifies a number of necessary conditions for adoption (and diffusion) of a technique. An actor, e.g. a firm, would need to have information about, access to and a subjective interest in adopting the technique. That firm must also be organized to adopt and use the technique, have the power to adopt it and the knowledge of how to use it. Under the assumption that a firm needs to make profits, its subjective interest in adopting a given technique can largely be seen as dependent upon the characteristics of the technique in relation to other techniques and to the structural environment, e.g. the relative factor prices, of the firm. The decision to adopt a technique or not can also be seen as being dependent upon the character of the actor. In section 2.2.3 larger firms were suggested to be better equipped to adopt new techniques than smaller ones, partly on account of larger risk taking capacity and better access to information. Over time, we suggested that two important determinants of the rate of diffusion of flexible automation techniques were the changing behaviour of the supplying industry and the diffusion of information about the technique and its advantages. Let us, in terms of this framework, discuss the reasons for the unequal rate of diffusion of flexible automation techniques between the developed and the developing countries.

The most obvious reason for non-adoption of flexible automation techniques in the developing countries is that the characteristics of the technique and the structural environment of the actors together do *not* create a subjective interest of the actors to adopt the technique. For example, the poor use of robots in Korea can to a considerable extent be explained by the relative factor prices in Korea and the technical characteristics of robots in relation to manual labour. However, as was shown in chapter 10, there are also other reasons for not choosing robots in the Korean structural context. In one particular case, the lack of application engineering skills in Korea in conjunction with a refusal by the supplying firm in Japan to send scarce application engineers to Korea was the main obstacle to adoption. Thus, lack of knowledge about how to use the technique and the behaviour of the supplying industry became important obstacles to the diffusion of robots.[7]

For NCMTs and CAD, we would argue that factors other than those referring to the technical characteristics of the technique and the structural environment – e.g. lack of information or skill shortages – are very important or even dominating in explaining the present relatively low level of diffusion in the developing countries. This means that the present low level of diffusion of NCMTs and CAD would not fully, or perhaps not even mainly, be a function of a lack of potential profitability from adopting the techniques. Instead, it would be an effect of other factors which can be altered through a proper government policy. To the extent that these obstacles to diffusion could be mitigated, the level of diffusion of flexible automation in the developing countries could increase, and consequently some of the negative effects experienced now in terms of reduced international competitiveness could be negated.

In particular, we would emphasize that CAD, together with NCMTs, is the most interesting of the four flexible automation techniques from the point of view of the developing countries. A basic reason for this is that in most developing countries design capability is underdeveloped in relation to production capability. Design and draughting experience is a scarce resource in most developing countries. This is indicated by their heavy reliance on foreign technical licences. Through two mechanisms CAD can help firms in developing countries to overcome a shortage of design engineers. First, CAD software embodies accumulated design and draughting experience which, through the CAD unit, can be put at the disposal of engineers based in the developing country. In other words, CAD may be used for leap-frogging in the field of design in developing countries. Secondly, the productivity of the design engineers can increase significantly with the use of CAD, which means that fewer engineers are needed to finish a given design. CAD is thus a skill-saving technique.

These productivity increases apply to basic design work, where CAD helps the experienced design engineer increase his/her productivity through eliminating the relatively simple and time-consuming draughting elements of his/her work (Kaplinsky, 1983). For detailed design work, the productivity increases can be very large. This can be of special interest to firms in developing countries which buy technical licences from abroad and often need to modify these slightly in response to the local market and to match raw material and component availability. In addition, CAD is a necessary tool for some types of design, particularly in the field of electronics. Finally, subcontractors, in the developed or the developing countries, must adopt CAD if the buyers of the components specify the product in CAD terms – which is beginning to be the case in some OECD countries.

Hence, there are very strong reasons for the developing countries to adopt CAD. It has also become easier to adopt CAD since the development of personal computer-based CAD units since these are easier to master and their low cost has also made it possible for smaller firms in the developed countries to adopt CAD. Hence CAD has a large potential for diffusion in the developing countries and as was pointed out in section 12.1, there has indeed been a rapid diffusion of CAD in the NICs in the past years.

The prime characteristic of NCMTs in relation to manual machine tools is that they save on skills among the operators. In addition, the productivity per operator increases significantly. Jointly, this means that the operator skill content per unit of output declines greatly with the use of NCMTs. To this we need add, however, the possibility that the greater skill content of the repair and maintenance staff for NCMTs reduces the extent of skill-saving associated with NCMTs. Furthermore, it has to be emphasized that certain skills, e.g. setting and repair skills, are *necessary* for the proper use of NCMTs. But the number of individuals needed to hold these skills is small in relation to the number of skilled machine tool operators required with the use of conventional machine tools.

In conclusion then, the skill-saving associated with the use of CAD and NCMTs is significant. This provides a strong argument in favour of a further diffusion of these techniques in the developed countries. The case for NCMTs may sometimes be strengthened, as we argued in section 3.4, by the capital-saving nature of NCMTs in some applications. However, capital-saving is probably not a general feature of the use of NCMTs.

Which then are the obstacles to the diffusion of, particularly, CAD and NCMTs? In previous chapters we have argued that explanations for a slow diffusion of a new technique, apart from the characteristics of the technique and the structural environment of the (non) adopter, may lie in:

- a lack of information about the technique;
- a lack of knowledge about how to use the technique;
- the behaviour of supplying industry.

In chapter 2, we emphasized that the diffusion of information about new techniques may be very slow. We also suggested that this obstacle to adoption is especially important for the developing countries and, above all, for firms that are not subsidiaries of multinational enterprises. The task of providing information about new techniques to, primarily, national firms in the developing countries is thus very important. A number of developing countries, e.g. India, Korea and Taiwan, have created national institutes whose functions are partly to teach local entrepreneurs about the advantages of new techniques. Apart from these activities, we would suggest that the experiences of the developed countries in setting up 'show cases' could be evaluated. In this way, private firms may subsidize their investment in a new technique, e.g. a CAD. In return, they are expected to permit representatives from other firms to study closely and learn from their investment.

As far as knowledge and skills are concerned, this is basically a question of the proper functioning of the education system. We would emphasize that the technical training schools must reorient their curricula so that relevant human skills are available. More specifically, this means, in the case of NCMTs, that less emphasis should be put on educating traditional machine tool operators and more emphasis be given to educating operators of NCMTs and the associated programming, setting and maintenance staff. In short, a limited number of engineers and technicians with a knowledge of, inter alia, electronics, should be educated instead of a considerably larger number of skilled operators of manual machine tools.

In the longer run, additional changes are needed. The skill requirements associated with FMSs are rather different from those of stand-alone NCMTs. The latter technique demands fairly specialized skills at a rather low level whilst the former demands maintenance and repair personnel who are skilled in a number of fields, e.g. mechanics, electronics, hydraulics etc. The educational system must therefore be ready to provide skills which are not of a traditional type. In terms of the design engineers educated, these will need to be taught not only how to design with CAD, but also how to modify CAD software.

As far as the behaviour of the supplying industry is concerned, we argued in section 9.2.3 that the evidence suggests that a local NCMT industry in the NICs normally provides higher priced items and in less variety than are available on the international market. Naturally rather than being an unavoidable state of affairs, this is rather a result of the way in which the local industry is fostered. The local industry may, however, also have a positive effect on the diffusion, since local firms appear to be more inclined than foreign ones to teach local customers about the technique, and supply good repair and maintenance services.

Indeed, the behaviour of the international industry in this respect provides a strong argument in favour of fostering actors that supply these functions, e.g. local firms producing, for example, NCMTs or CAD software. The *way* in which these firms are fostered, however, is crucial for the rate of diffusion. We noted in the discussion of access in section 9.2.3 that the local industry provides a larger share of the local demand for NCMTs in some NICs than is the case in the USA or the UK. This is to a considerable extent a result of fostering of the local industry through applying quantitative import restrictions. Another possibility is subsidized credit to local firms producing the techniques – but not combined with import restrictions – while the industry is in its infancy to ensure that local prices are not above international prices and that local industry has access to the whole range of machines available in the international market. To the extent that foreign suppliers are permitted to sell in the local markets – be it either CAD, NCMTs, robots or FMSs – the government needs to make sure that service follows sales through legislative or other measures.

In summary, many developing countries, i.e. Group A and Group B countries and some Group C countries – should support the diffusion of CAD and NCMTs because of their skill-saving nature. Such support demands educational efforts as well as government measures to build up, or strengthen, maintenance and repair capabilities. For a few (Group A) countries it may also mean support to a local industry producing the techniques.

Industrial robots and FMSs are less relevant for the developing countries; robots because of their factor-saving bias and FMSs because they are immature and complex. In other words, there are at present no strong reasons for governments to support their diffusion in developing countries, except in a few semi-industrialized countries such as Singapore and Korea, i.e. countries with a high technological capability and with fairly high wage levels.[8]

Notes

1 We would also underline that if and when NCMTs and CAD are diffused to a greater extent in the developing countries, the skill-saving character of these techniques can be quite beneficial to investors in these countries. Indeed, to the extent that skilled workers and design engineers are more scarce in the developing countries than in the developed countries, NCMTs and CAD can be *more* advantageous to investors in the developing countries than to investors in the developed countries (see Jacobsson, 1982, for the case of NCMTs).

2 To the extent that Group C (and B) countries want to enter into production of such capital goods, which are greatly affected by flexible automation techniques, the adoption of these in the OECD countries and the NICs means that price competition has become fiercer and the demands for short lead times greater. The competitive environment has thus become tougher for new entrants. However, the skill-saving effects of NCMTs and CAD can give some advantages to Group B and C countries adopting the new techniques. As is discussed further in the text, the developing countries need to adjust their educational system in order to benefit from these skill-saving effects. Furthermore, the impact of flexible automation techniques on the competitiveness of the developing countries would appear to be different depending on what the alternative to flexible automation is. If the alternative is manually operated, stand-alone machines such as engine lathes instead of NCMTs, the impact on the developing countries appears to be negative. However, if the traditional alternative to flexible automation is fixed automation in the form of transfer lines, the picture alters significantly. Transfer lines are normally used for the production of long series of homogeneous products and are frequently used in the automobile industry. Fixed investment is high and large scale production is needed to justify this investment. Generally

speaking, investments of this type are less justified in the developing countries than in the developed countries given the more limited market in the former. The move towards flexible automation observed in the developed countries, as shown, for example, in appendix tables 5.1 and 5.2, opens up a very interesting alternatives for those developing countries that have a smaller local market. These countries now have the opportunity to start producing products, such as diesel engines, with flexible automation, and compete with imported products at much lower volumes of production than was possible if the technique used was transfer lines.

3   There is a risk that those Group A countries that do not follow suit in the diffusion of flexible automation techniques find themselves in the category of losers. This may be a large risk for all Group A countries with regard to products where robots and FMSs are used in the OECD countries, since these techniques are not widely diffused in the Group A countries. (In the case of robots and FMSs, Singapore seems to be an exception in this respect.)

4   The product groups included were ISIC 3811, 3822, 3823, 3824, 3829, 3831, 3832 and 3843. In the Swedish case, for which data are available, we have also included 3845. The lower of the two figures for Korea results from including only parts of ISIC 3832, which is 'manufacture of radio, television and communication equipment and apparatus'. In the calculation resulting in the lower percentage we have included only ISIC 38322 which is 'manufacture of line communication equipment' and ISIC 38325 which is 'manufacture of electronic tubes and other electronic parts and components'. Unfortunately, the Swedish data are not detailed enough to enable us to make an equivalent exclusion. In addition, the Swedish data include 3845. For India problems arise from the fact that Indian industrial classification is different from the ISIC one. In the Indian figures we have included the following groups: 343, 350, 351, 353, 354, 356, 357, 359, 360, 367, 374. The Swedish figures are from 1983 and the sources are Swedish Central Bureau of Statistics (1983). The Korean data are from 1978 and the source is Economic Planning Board, 1984. The Indian data are from 1978 and the source is Edquist and Jacobsson, 1982.

5   With the exception of SITC 7293, the five product groups that had the highest share of the manufacturing exports of the developing countries in 1983 were: SITC 7353, ships and boats, other than warships (3.66 per cent); SITC 7359, special purpose ships and boats (1.74 per cent); SITC 7242, radio broadcast receivers (1.63 per cent); SITC 7249, telecommunications equipment n.e.c. (1.62 per cent); SITC 7222, apparatus for electrical circuits (1.25 per cent). (Source: UNCTAD trade data bank.)

6   See Schumacher (1983) for an analysis of the determinants of developing countries' exports to the Federal Republic of Germany.

7   FMSs, compared to stand-alone NCMTs, are basically a labour-saving innovation, where an FMS saves on both direct and indirect labour. Given this feature and the present immaturity of the technique, which means that it is very engineering-intensive, we believe that it will not diffuse quickly in the developing countries.

8   FMSs may also decrease the minimum efficient scale of production – as compared to fixed automation – for certain products. This may make them interesting for the developing countries, since the sizes of their domestic markets are limited for many products (see n. 2). If this is the case, an adequate reorientation of the educational system is needed.

Statistical Appendix to Chapter 12

Appendix table 12.1 Export performance of the developing countries in those product groups judged to be mostly affected by flexible automation

SITC group		(1)[a]	(2)[a]	(3)[a]
7293	Thermionic valves etc.	18.4	50	3.15
7221	Electric power machinery	7.0	136	0.67
7118	Engines n.e.c.	6.5	83	0.05
7115	Internal combustion engines, not aircraft	4.2	50	0.69
7173 7191 7193 7196 7197 7198 7199	Other non-electrical machinery[b]	4.6	61	1.35
7341	Aircraft	3.7	200	0.33
7184	Construction and mining machinery	3.4	27	0.44
7171 7172 7181 7182 7183 7185	Special industrial machinery excluding construction and mining machinery[c]	2.9	67	0.21
7328	Bodies and parts motor vehicles, excl. motorcycles	2.8	7	0.51
7125	Tractors	2.8	163	0.14
7192	Pumps and centrifuges	2.5	79	0.35
7349	Parts of aircraft	2.2	6	0.23
7151	Machine tools for working metals	1.6	89	0.11

[a] (1) Developing countries' share of world export in 1983; (2) the percentage by which developing countries' export growth exceeds the growth in exports for the world in the period 1970–83; (3) the share of the various product groups in all manufacturing exports from the developing countries in 1983.
[b] The composition of these SITC groupings roughly corresponds to ISIC groupings 3824 and 3929.
Source: UNCTAD trade data bank.

Appendix table 12.2   Top ten SITC four digit product groups
(engineering industry) as regards the share of the developing countries
in world exports, 1983

	SITC group	Share in world exports (%)
7359	Special purpose ships and boats	33.7
7242	Radio broadcast receivers	31.2
7316	Railroad and tramway freight cars, not mechanically propelled	26.2
7353	Ships and boats, other than warships	21.7
7241	Television broadcast receivers	20.8
7293	Thermionic valves etc.	18.4
7358	Ships, boats and other vessels for breaking up	13.2
7250	Domestic electrical equipment	12.0
7315	Railroad and tramway passenger cars, not mechanically propelled	9.0
7291	Batteries and accumulators	8.9

Source: UNCTAD trade data bank.

PART IV
Main Findings

13
Review and Conclusions

The engineering industry is undergoing a fundamental technological transformation. This transformation is taking place in all spheres of engineering enterprise, including design and production. The objectives of this book have been to describe and analyse the process of diffusion of four techniques which are all central to the development of the 'factory of the future'. These techniques are numerically controlled machine tools (NCMTs), robots, flexible manufacturing systems (FMSs) and computer aided design (CAD) systems. The book has covered not only the diffusion of these techniques in the advanced OECD countries, such as Japan and Sweden, but also in the so-called newly industrializing countries (NICs). The book thus attempts a global analysis of the diffusion of new techniques. One objective has been to describe the process of diffusion and to discuss the determinants and obstacles to this diffusion. Another important objective has been to address some consequences of this diffusion. For the advanced OECD countries we have tried to contribute to the debate on the impact of new techniques on employment and international competitiveness. For the NICs, our concern, apart from description and analysis of the diffusion of the new techniques, has been to analyse to what extent this diffusion affects the competitiveness and the industrialization prospects of these countries.

The book has been divided into three parts and we will now summarize the main findings of each.

13.1 Part I Introduction

Apart from a more general introduction to the book, Part I presents a conceptual framework for our study on the diffusion of the new flexible automation techniques. We distinguish between three different, although overlapping, approaches to the diffusion of techniques. The first combines a structural and an actor-oriented approach and proposes that three different determinants be focused upon, namely:

1 The character of the new technique in relation to the old technique, e.g. whether it saves on labour or/and capital etc.
2 The structural environment of the (non)adopter, e.g. the factor prices prevailing and the infrastructural conditions existing in the society.
3 The character of the actors as such, e.g. the size and other characteristics of the (non)adopter.

The second approach looks at the diffusion process in a dynamic way and identifies three different stages in the diffusion process, namely, the introductory phase when the technique is first used, the growth phase when the rate of diffusion of the technique accelerates greatly, and the maturity phase when the technique has reached its potential diffusion level and the situation is stable until a yet newer technique begins to substitute for the now older technique. Often it is assumed that the diffusion over time takes the shape of an S-curve. This approach identifies as key determinants of the movements of the various techniques along the S-curves the behaviour of the supplying industry and the diffusion of information.

The third approach is more general and incorporates many of the elements of the other two approaches. The concept of 'social carriers of techniques' identifies a number of necessary conditions for adoption and diffusion of a technique. An actor, e.g. a firm, needs to have a subject interest in choosing the technique, to be organized to be able to take the decision, to have the power to materialize its interest, to have information about the technique, have access to the technique and have the knowledge of how to handle it. If these conditions are fulfilled, the actor is a social carrier of a technique and the technique is adopted. At the same time, if one of the conditions is not fulfilled, the technique is not introduced and the unfulfilled condition can be seen as an obstacle to the diffusion of the technique. A government policy can therefore identify unfulfilled conditions as points of intervention.

In the subsequent analyses, we use a mixture of these approaches, depending on the objectives of the sections. In the second part of the book, dealing with the diffusion of flexible automation techniques in the OECD countries, our objective is primarily to describe the diffusion of these techniques and to assess their degree of maturity. Hence, the second approach dominates but we use the first also. In the section dealing with the NICs, where the degree of maturity of the techniques is given, our objectives include a somewhat more ambitious discussion of the determinants of the diffusion of the techniques, including a discussion of the reasons behind the slower diffusion in the NICs than in the OECD countries. Emphasis is therefore given to the first and third approaches.

13.2 Part II The Diffusion of Flexible Automation Techniques in the Engineering Industries of the OECD Countries

In Part II we describe the four techniques and their diffusion. We also discuss the impact of this diffusion on employment and international competitiveness.

13.2.1 NUMERICALLY CONTROLLED MACHINE TOOLS

The bulk of production in the engineering sector is accounted for by small and medium batches. Flexible machine tools are therefore required. Numerically controlled machine tools (NCMTs) represent a flexible technique that is also automated, in contrast to flexible, manually operated machine tools, e.g. engine lathes. NCMTs were invented in the 1950s, but did not begin to be diffused on a large scale until the mid 1970s. For example, in Sweden, the share of NCMTs in total machine tool investment rose from 26 per cent in 1987 to 59 per cent in 1984.

However, the use of numerical control varies greatly between different types of machine tools. It is primarily metal-cutting machine tools, as distinguished from metal-forming machine tools, that are equipped with numerical control units. Furthermore, among the metal-cutting machine tools it is primarily those machine tools that perform milling, drilling, boring and turning functions that have been transformed into NCMTs. In the leading OECD countries, 76 per cent of the value of production of those machine tools performing these functions were NCMTs in 1984. For grinding and finishing machines, however, the corresponding figure was only 11 per cent, i.e. there are also considerable differences within the group of metal-cutting machine tools.

NCMTs, especially those performing the above-mentioned functions, went through their growth phases – in the S-curve terminology – in the mid and late 1970s, and are now in the maturity phase. This is reflected in the size distribution of those firms that use NCMTs, where smaller and medium-size firms dominate in both Japan and the USA.

The process of maturation of the NCMTs technique was intimately connected to the behaviour of the supplying industry. It was primarily the behaviour of some leading Japanese firms which led to a global restructuring of the NCMT industry. They successfully targeted on the needs of the smaller and medium-size firms, which demanded a slightly different technique than was hitherto available in the market. This led the industry to, and beyond, the point of maximum curvature on the S-curve.

The industrial distribution of NCMTs reflects quite closely the importance of machining in the various stages of the production process. Other considerations, such as the demand for flexibility or precision, also influence the extent of use of NCMTs. Among the large number of branches in the engineering sector, it is the following which show the greatest use of NCMTs (in the USA):

- Construction, mining and material handling equipment.
- Miscellaneous machinery, except electrical.
- Metal-working machinery.
- Miscellaneous transport equipment.
- Engines and turbines.
- General industrial machinery.
- Special industrial machinery.

As would be expected from a mature technique, the economic efficiency is well established. Given that machining plays a role in the production process, the use of NCMTs is basically seen as a way to reduce costs per unit of output. It is clear that labour productivity normally increases greatly with NCMTs. Capital productivity may also increase in some cases. A further important aspect of the use of NCMTs is that it saves greatly on skilled labour per unit of output.

13.2.2 INDUSTRIAL ROBOTS

Robots began to be diffused in the 1970s. In the period 1974–84 the average annual rate of increase in the stock of robots was 44 per cent in the major OECD countries. The robot technique could be said to have entered into its growth phase around 1980, where it clearly is now. In 1984 the global investment in robots amounted to between US$1,200 and 1,300 million, out of which Japan accounted for around US$800 million.

In 1984 around 100,000 robots existed in the OECD countries. Of these, 65 per cent were installed in Japan, which also leads in the 'density' in use of robots. In 1984 Japan had 113 robots per 10,000 employees in the engineering industry, whilst Sweden had 70 and Belgium 28. The UK was at the bottom of the league with nine robots per 10,000 employees in the engineering industry.

Robots are divided into three categories – handling, process and assembly robots. Process robots are currently the dominant type and among these welding robots account for the majority. Handling robots comprise the second most important category.

Hitherto, assembly robots have not diffused to any significant degree. Therefore one of the most interesting features of robot diffusion is the growth in the proportion of assembly robots. In the Federal Republic of

Germany, for example, this proportion increased from 4 per cent of the stock of robots in 1980 to 7 per cent in 1984. During 1985, 301 assembly robots were installed in the FRG, i.e. the stock increased by 67 per cent to 753 units. This development is of great interest, given the fact that a very large share of the total hours worked in the engineering sector is spent on assembly operations. Until recently, this task has mainly been carried out manually, but, for example, the assembly of windscreen wipers at one Electrolux factory in Sweden is now being carried out by robots.

It was the automobile industry which began to use robots on a larger scale and it is still the dominating user industry. However, other sectors, notably the electrical and non-electrical machinery industries, are now adopting robots also. With the exception of the UK, however, it seems as if the larger firms are still dominating over smaller and medium-sized ones in the use of robots.

The adoption of robots is, in the short run, mainly motivated by reduced costs for unskilled and semi-skilled labour. Hazardous work environments can also be a contributing factor. In the longer perspective, however, firms often buy robots in order to upgrade the technological level of the enterprise. The benefits of the investment lie therefore partly in knowledge generation, which is expected to become a benefit, in narrow economic terms, only in the more distant future. Investment in robots is thus sometimes considered to be a strategic investment, motivated by the role robots play in the development of more advanced manufacturing systems.

13.2.3 FLEXIBLE MANUFACTURING SYSTEMS

The diffusion of stand-alone NCMTs has now become rapid. The main development in the past few years has been the introduction of production systems which are both flexible and automated. As an overall term for these systems, flexible manufacturing systems (FMSs) is normally used. However the term covers systems of different sizes and with different degrees of automation. In this book we distinguish between flexible manufacturing modules (FMMs), flexible manufacturing cells (FMCs) and flexible manufacturing systems proper (FMSs). FMMs are the smallest systems, comprising one NCMT with an automated material handling unit and a monitoring system. FMCs are systems with at least two NCMTs whilst FMSs proper are larger systems. A FMM is normally chosen when the number of variants is great and each is produced in smaller batches. A FMS proper is normally used when the number of different parts is smaller, e.g. 10–100, but each variant is produced in larger batches.

The diffusion of FMMs, FMCs and FMSs is of relatively recent origin. All the systems are probably in their introductory stages. This applies in particular to FMSs proper. In 1984–5 it was reported that around 300 FMSs proper existed in the world: they incorporated 2,139 machine tools. This is less than 1 per cent of the global stock of NCMTs. FMMs and FMCs however, are probably close to their growth phases, globally speaking. In the case of Sweden these techniques have probably already entered into their growth phase. In a special study of the case of Sweden, we estimate that of all the NCMTs installed during 1984–5, around 11 per cent were part of FMMs and 14–18 per cent were part of FMCs whilst 2 per cent were part of FMSs. Available evidence strongly suggests, however, that the rate of diffusion of systems is far higher in Sweden than in other European countries.

The available information on the industrial distribution is limited to FMSs proper. Which products then are feeling the earliest impact of this new technique? At the three digit ISIC level, it is clearly the non-electrical machinery industry and the transport equipment industry which account for the bulk of the FMSs proper. In terms of final products which incorporate parts produced in an FMS, the main impact is in the following areas:

- Automobiles and trucks.
- Machine tools.
- Tractors and construction machinery.
- Aerospace.
- Diesel engines.
- Electric motors.
- Pumps, valves and compressors.

Together these products account for approximately three-quarters of the installed FMSs proper globally. Of course not all components in these products are produced with FMSs. It is mainly various types of housing, e.g. gearbox housing or transmission cases, shafts, cranks and axles, gear parts, cylinderheads, columns, heads and beds for machine tools and engine frames which are produced in FMSs.

Generally speaking, there are four aims in introducing various types of FMS, including FMM and FMC. These are:

- Improved machine utilization.
- Reduced cost for work in progress and stocks.
- Increased labour productivity.
- Ability to react faster to changing market conditions.

There are several ways to attain these goals, one of which is to introduce new techniques. The evidence on the economic efficiency of the various types of systems varies in availability. A number of advantages are clearly there for users of FMMs, notably an increase in the utilization of the machines and an increase in labour productivity. A FMM is a relatively uncomplicated technique and this is a further advantage to the user. As far as FMCs are concerned, it is claimed that the payback period for an FMC usually ranges from two to four years, which is quite a normal payback period for machine investments. This is a more complicated technique, however, and often a lot of engineering effort needs to be put in by the user in the running-in stage. Finally, FMSs proper have until recently been characterized as mainly experimental. The evidence tends to suggest, however, that real economic benefits are accruing to users in terms of both reduction in unit costs and a decrease in lead times.

13.2.4 COMPUTER AIDED DESIGN

Computer aided design (CAD) dates back to the end of the 1960s when mainframe computer-based design systems began to be marketed. With this technique the designer can communicate with a computer which serves as an electronic draughting board. Later, minicomputer-based systems became available and in the early 1980s CAD systems based on personal computers (PCs) emerged.

The most remarkable development of CAD in the 1980s is clearly the appearance and growth sales of PC-based systems. The example of Sweden can illustrate the development. In 1982 there were 205 mainframe- or minicomputer-based CAD systems in Sweden. This was before the PC-based units were sold in that country. Assuming that on average each of these larger systems had four workstations connected to it, this would mean that around 800 workstations existed for CAD work in 1982. From 1983 to September 1985 one supplier alone sold 750 PC-based systems in Sweden, that is, almost as many workstations as the total stock in 1982. Hence, the emergence of PC-based systems led to a very rapid diffusion of CAD in Sweden. This applies also to the global situation and it is estimated that in 1985 around one-third of the CAD workstations available were based on PCs. With the emergence of PC-based systems, it is clear that CAD is well into its growth phase. This is also evidenced by the fact that whilst the larger CAD systems were almost exclusively sold to large firms, PC-based systems are also being adopted on a large scale by smaller firms. In the Swedish case, about half of the PC-based CAD units were sold to smaller firms in 1984.

As far as the sector-wise distribution is concerned, the electronics industry is the single most important user both of the larger and of the PC-based systems.

The most common reasons for introducing CAD are:

- Improved productivity for designers and draughtsmen.
- Shortening the lead time from conception to production and from order to delivery.
- Performing work which is too complex for manual design and drawing.

The motives for introducing CAD vary however between industries. In the electronics industry, the complexity of the product, e.g. integrated circuits, makes the use of CAD imperative. In the automobile and aerospace industry, a dominating motive lies in shortening the lead time from design to production and in perfecting the product design. In the mechanical machinery industries an important reason for the use of CAD lies in the rationalization potential in situations where a given basic design is used for a range of machines, e.g. different sizes of paper machines or pumps. This type of application is now growing in the mechanical engineering sector.

13.2.5 FLEXIBLE AUTOMATION AND INTERNATIONAL COMPETITIVENESS

Having addressed the process of diffusion of the four flexible automation techniques, we proceed to discuss the effects of this diffusion on international competitiveness among the advanced OECD countries. The impact is first analysed through assessing the quantitative importance of robots and NCMTs in investments in machinery and equipment in some advanced OECD countries. With the exception of the USA, the trend is clearly a sharply rising share for these two techniques in investment in machinery and equipment.

The sector-wise distribution of the stock of flexible automation techniques is then focused upon. The metal product sector (ISIC 381) is badly represented, whilst the remaining sectors have a fair share of several of the techniques. CAD, however, is mainly used in the electrical machinery industry (ISIC 383) whilst FMS is mainly used in the non-electrical machinery industry (ISIC 382) and in the transport machinery industry (ISIC 384), which also dominates the use of robots.

In terms of product groups affected by flexible automation, we identify the following as the prime ones:

- Automobiles.
- Heavy electrical equipment.
- Aircraft and parts.
- Cutting tools, pumps, valves and compressors.
- Construction machinery, tractors and machine tools.
- Special industrial machinery.
- Electronic components.

In terms of the impact at firm level, two case studies are reported on firms producing submersible pumps and diesel engines. The case studies clearly show the importance of the new techniques in these firms. The technological transformation that these firms are undergoing results in a reduction in unit costs and a reduction in lead times. The case studies also show that a range of factors other than new techniques need to be considered when analysing the competitiveness of these firms. However, although the cost efficiency of the plants is certainly not the only factor determining the competitive strength of firms, one can, on fairly safe grounds, suggest that the profit level is quite sensitive to the introduction of the new techniques and to the skill with which they are applied. A positive effect on the profit level of a firm can well be of great importance to its long-term survival and development.

It then becomes of interest to compare the level of diffusion of the new techniques between the various advanced OECD countries. We calculate 'density' measure which comprises the number of units of each technique installed divided by millions of employees in the engineering industry. A comparison is then made for the four techniques for the Federal Republic of Germany, Japan, Sweden, the UK and the USA. Two main results of the comparison are:

1 Among these countries, with the exception of CAD, Japan and Sweden have much greater densities than the other countries.
2 There is a considerable discrepancy between the four techniques as regards the size of the difference in densities, where the greatest discrepancy occurs for the less mature techniques. For NCMTs the density varied in 1984 by a factor of 2.1. For CAD, robots and FMSs the factor was 5.1, 14.5 and 18.3 respectively. There is thus a very considerable difference in the density in use of these techniques.

To illustrate the length of the period during which early adopters can make 'super-normal' profits on account of an early introduction of these new techniques, the diffusion process in very similar countries (Sweden, Federal Republic of Germany, the UK and the USA) were studied. In 1967, a discrepancy in density of NCMTs between the slowest diffuser, the FRG, and the leading nation, the USA, of a factor of 9.3 was recorded, whilst the discrepancy between the leading and the slowest diffuser was reduced to a factor of 4.7 in 1976 and finally to a factor of 2.1 in 1984. Thus, it took 17 years for the national differences in densities to be reduced to a maximum of a factor of 2.1, which is still, however, a surprisingly high figure. As for NCMTs, we expect that the discrepancy in densities for the other techniques will be reduced over time. However, as the example of NCMTs illustrates, the length of the period in which the early adopters can earn 'super-normal' profits may be quite long – between one and two decades. However, as mentioned above, the use of new production and design techniques is only one determinant out of many of the international competitiveness of firms and countries.

13.2.6 FLEXIBLE AUTOMATION AND EMPLOYMENT

The relation between technical change and employment is a highly complex issue that can be discussed at a number of levels, e.g. machine, firm, sector, and country. Technical change can, furthermore, affect both the quantitative and the qualitative sides of employment. Finally, there are a number of determinants other than technical change that influence employment.

In terms of the *quantitative* effects on employment at the machine level, it is clear that flexible automation techniques have a substantial labour replacement effect. There are less opportunities for work with the new techniques than there would have been with the old, given the same output level. Taking the cases of NCMTs and robots jointly, an equivalent of between 2.3 per cent (UK) and 6.9 per cent (Japan) of the workforce in the engineering sector have been replaced. This estimate is, for a range of reasons, very conservative.

However, the fact that Japan has the lowest and the UK the highest rate of unemployment illustrates that it is impossible to aggregate the effects at the machine level to the national level. It is indeed not possible to make such an aggregation even to the firm level. At the firm level, compensationary effects may operate, e.g. through an increased demand due to, for example, improved competitiveness. At the national level, the general economic policies of governments strongly influence the demand for labour, as do attitudes to unemployment and associated policies. Political priorities often act as a powerful determinant of unemployment, which we illustrate in our discussion of Cuba after the revolution in 1959.

In terms of the *qualitative* effects of technical change on employment, NCMTs are generally speaking associated with a de-skilling of the operators. At the same time, however, a number of highly skilled jobs are associated with the programming, setting, repair and maintenance of the NCMT. In spite of this, a substantial reduction in the skill content per unit of output is linked to the use of NCMTs. The de-skilling of the operator, however, is not solely or fully determined by the technique as such. There are many ways of using NCMTs, including the case where the operator is a very skilled worker who himself does the programming and setting as well as other less qualified tasks. Another possibility, which appears to be the most common one, is that the more qualified tasks are carried out by engineers and technicians, leaving only the tasks requiring little qualifications to the operator. Owing to the de-skilling of the operators and an improved labour productivity, the use of NCMTs is thus associated with a large decrease in the skill content per unit of output.

In the case of robots, those employees which are replaced are normally unskilled or semi-skilled workers, whilst of those jobs that are created (e.g. in maintenance and repair) more than 50 per cent are highly skilled.

Consequently, the introduction of robots is on the whole positive in terms of skill requirements and quality of jobs.

In the case of FMS, it is even more pronounced that the new jobs created are of a more qualified nature than those that they replace. This applies both to the installation and running-in phases and to the operation, repair and maintenance aspects. Often the skills required are mixtures of several different kinds of skill, such as electronics, mechanical engineering and hydraulics. In summary, as far as robots and FMSs are concerned it is clear that fewer people are required per unit of output, and that these few need to have higher qualifications than the previous jobs required.

13.3 Part III The Diffusion of Flexible Automation Techniques in the Engineering Industries of the Newly Industrializing Countries

In Part III we describe the diffusion of the flexible automation techniques in the NICs, address the determinants of the level of diffusion and discuss the consequences of this diffusion for international competitiveness of the NICs. The latter discussion is presented in the final chapter (chapter 12) of Part III, where we also discuss policy implications for the developing countries.

13.3.1 NUMERICALLY CONTROLLED MACHINE TOOLS

The stock of NCMTs in the NICs varies between 500 (Argentina, 1985) and 2,680 (Korea, 1985). In terms of the share of NCMTs in total investment in machine tools, the NICs vary between 7 and 23 per cent, whilst the range in the OECD countries was between 40 and 62 per cent in 1984. Hence, the NICs are far behind in the diffusion of NCMTs.

Those NCMTs which are diffused in the NICs are distributed sector-wise in broadly the same way as in the OECD countries, with the non-electrical machinery industry (ISIC 382) and the transport machinery industry (ISIC 384) as the dominating sectors. As would be expected, given the relatively slow diffusion, the larger firms dominate in the use of NCMTs. The exception appears to be Argentina, where smaller and medium-sized firms are adopting the new technique to a considerable extent.

Which then are the determinants of diffusion of NCMTs in the NICs? In the following discussion on NCMTs we use the concept of 'social carriers of techniques' to provide the structure of the discussion. (The same applies to the subsequent sections dealing with the diffusion of CAD and robots/ FMS in the NICs.)

Of the six conditions defining a social carrier of technique, we do not deal with conditions 2 (organization) and 3 (power), as these are normally automatically fulfilled by firms in capitalist countries. Instead, we refer to conditions 4 to 6 (information, access and knowledge) and to condition 1 (interest).

From a range of studies undertaken in the NICs, it is clear that substantial problems have existed and still exist as regards the diffusion of information about NCMTs. This was also the case in Europe a decade ago. This is particularly a problem for the smaller and medium-sized firms which adopt NCMTs slower than do the larger firms.

Access to the new technique is often more restricted in the NICs than in the developed countries on account of the import restrictions prevalent in many countries. These restrictions often have the effect that the import share of investment in NCMTs is higher in the OECD countries than in the NICs. This is very serious, since a key feature of the international NCMT industry is a considerable product differentiation. Firms in specific countries normally specialize in the production of a limited set of models/ sizes and imports cover the remaining domestic demand. With the exception of the Japanese industry, no OECD country can satisfy local users with all types of NCMT. Obviously, the low import share of investment in NCMTs in the NICs means that local firms' access to some versions of the new technique is restricted.

As far as knowledge about how to use and maintain NCMTs is concerned, available evidence points to problems in the supply of repair and maintenance service from the producers and a general lack of maintenance skills. Partly these problems occur as a consequence of suppliers' lack of interest in the relatively small NIC markets. This opens up the possibility that local suppliers can achieve a competitive strength on the basis of a better supply of these services.

Clearly, problems regarding information, access and knowledge retard the diffusion of NCMTs in the NICs, in particular for medium and small firms. However, would we expect firms in NICs to have a similar interest in adopting the techniques as OECD firms?

The choice of technique is in reality a very complex issue where a great number of determinants are at work. Only the more central ones are discussed in this book. In the literature, the opinion is often put forward that purely technical reasons, e.g. precision requirements, can partly explain the adoption of NCMTs in the NICs. Whilst it may well be so for a very limited number of products, we argue that the literature overrates this determinant on account of the fact that in the NICs the skill level among the conventional machine tool operators is often not adequate to produce components with high and consistent quality. In such instances, and if adequate programming, setting and maintenance skills are available, NCMTs can be chosen on account of the regular and high quality of their output.

Given that there is a choice between conventional machine tools and NCMTs, another important determinant of choice is factor prices. This refers to both the price of labour and the price of the machines as well as the interest rate. Generally speaking, NCMTs are applied when the batch to be produced is of medium size. However, if the labour costs are relatively low, the more labour-intensive conventional machines will be chosen to a greater extent. The same applies if the cost of the NCMTs is more inflated than the cost of conventional machines – as compared with world market prices. Naturally, this means that the protective policies *vis à vis* the local NCMT industry can have a profound effect on the interest of local firms in adopting this technique. It is also the case that NCMTs produced locally in the NICs are very costly in relation to the world market prices.

The expected capacity utilization is a further important determinant of the choice of technique. It may be that management perceives that it is easier to control capacity utilization with more automated techniques that reduce the control of the machine operators. This is not an unusual factor behind the choice of NCMTs in India.

Finally, as mentioned above, NCMTs are normally used when medium-sized batches are to be produced, although the application areas of NCMTs are currently being expanded to both smaller and larger batches. The type of market served by the potential adopter is therefore a further determinant of the choice between conventional machines and NCMTs. To the extent that the NICs produce more simple products in larger series, the potential for adopting NCMTs in the NICs would be less than in the OECD countries. Whilst this may be true, it may also be the case that the predominantly inward-looking NICs, at least when it comes to the engineering industry, require more flexible equipment than their OECD counterparts, since the local market for individual products is relatively small in the NICs. This would mean that in the NICs there may sometimes be a choice of NCMTs instead of more inflexible automatic machines which are used in OECD countries when very large batches are produced.

13.3.2 INDUSTRIAL ROBOTS AND FLEXIBLE MANUFACTURING SYSTEMS

The diffusion of robots in the NICs is still quite limited, with the exception of Singapore. It is the leading adopter among the NICs and has a density of use which is greater than that of the Federal Republic of Germany, the UK and the USA. An important obstacle to a wider diffusion of robots in the other NICs is the behaviour of the suppliers, which affects information about, access to and knowledge of how to use robots. A lack of local application engineering capability may be a particularly severe obstacle, which is further underlined by a reluctance on the part of foreign suppliers

208 *Main Findings*

to send scarce application engineers to marginal markets. Of course this, as in the case of NCMTs, opens up the possibility for local suppliers to compete. A fair number of the installed robots in the NICs are indeed locally made.

Robots substitute for unskilled or semi-skilled labour and the choice is very much a matter of straightforward capital–labour substitution. The low wages in the NICs do, of course, generally speaking reduce the interest in adopting the technique. For example, the annual cost of an employee in the transport machinery industry in 1981 was nearly four times higher in Japan than in Korea. However, for some specific applications, e.g. spot welding of automobiles, the payback period can be so short that the interest in adopting the new technique exists even in the NICs.

As far as FMS is concerned, the information varies in reliability between the countries. In Korea, in 1986 there were at least three FMSs proper (implying a density at the same level as the USA) and a number of smaller systems. The diffusion of the various types of systems has clearly begun in Korea and an interesting feature is that the early adopters are machine tool firms. One effect of this may be that there develops a local supply capability in Korea which can have important positive effects on the local diffusion of FMSs. As far as the other NICs are concerned, there may be FMSs in India and Singapore, although the systems referred to may be FMC. In Brazil, there is no FMS.

13.3.3 COMPUTER AIDED DESIGN

Computer aided design (CAD) systems are clearly being diffused to a considerable extent in the NICs. The stock of CAD seats varies from 74 (Argentina, 1983) to over 700 (in Korea, 1986 and India, 1987). In terms of the density in use of CAD, Korea and Singapore are the most advanced of the NICs, although most of Singapore's CAD systems are used outside the engineering industry.

As regards the determinants of the diffusion of CAD in the NICs, there are considerable problems with information, access and knowledge. Even in Korea, there is a general lack of information about CAD and, further-more, a lack of trained personnel to use CAD. The behaviour of the supplier also has a negative effect on the diffusion of CAD. There was, at least in 1984, a great deal of reluctance on the part of suppliers to get involved in the Korean market. In the Indian case the US government has applied an embargo, which virtually put an end to the diffusion of CAD until a local firm began to sell PC-based units.

The interest of firms in adopting CAD lies in several diverse factors. Naturally, the would-be adopter would need to undertake at least the detailed design activity within the firm in order to be interested in adopting

CAD. This condition is not fulfilled by a large number of firms in the NICs which rely on their licensor even for the detailed designs. However, once detailed or even basic design work is undertaken, what then are the reasons for adopting CAD?

For some firms, mainly in the electronics industry, the choice of CAD is technically determined. Today, the same reason could also be seen as applying to the design of chemical and other process plants. The ability to meet short lead times is also an important factor, particularly in the shipbuilding industry. A further important determinant is more specific to the NICs. In those countries, typically Korea, which are rapidly changing the structure of their industry and are experiencing a fast growth rate, a shortage of experienced design engineers arises. CAD can be very useful in such a situation for two reasons. First, CAD programs embody accumulated design and draughting experience which is then made available to the inexperienced designers. Secondly, CAD increases the productivity of the designers available, which means that a smaller number of experienced designers are needed per unit of output.

13.3.4 IMPLICATIONS FOR THE DEVELOPING COUNTRIES

It is clear that the NICs have a much lower level of diffusion of flexible automation techniques than do the OECD countries. Taking five OECD countries jointly (Federal Republic of Germany, Japan, Sweden, the UK and the USA) and comparing their densities in use of NCMTs, robots and CAD with those of five NICs (Brazil, India, Korea, Singapore and Yugoslavia), it is shown that for NCMTs the density in the OECD countries is 8.5 times that in the NICs; for CAD it is 8.3 and for robots it is 43.0. The data also reveal that the distance between the OECD countries and the NICs is the smallest for the most mature techniques, NCMTs and CAD. Still, however, the discrepancy is very great. Furthermore, in the case of NCMT, the NICs do not appear to be catching up with the OECD countries. They are far behind in terms of the share of investment in machine tools that is accounted for by NCMTs. Moreover, in terms of the rate of growth in the stock of NCMTs, they are not surpassing the OECD countries in any significant way. For CAD, however, the NICs are adopting the technique at a faster rate than the developed countries and the gap in densities has been reduced significantly in the past few years. Finally, there is also a difference between the NICs, with Singapore and Korea being far ahead of the other countries.

A general implication of the large discrepancies between the densities of the OECD countries and the NICs is that the former have clearly strengthened their competitive position *vis à vis* the latter. The advantages that the NICs have had in the form of cheaper semi-skilled labour are being eroded, in particular through the diffusion of robots and FMSs. Moreover, since these countries have adopted flexible automation techniques to a lesser extent than the OECD countries, they will benefit less from the productivity increases that adopters are able to reap through the use of the new techniques. However, the NICs can benefit and are benefiting from the skill-saving character of NCMTs and CAD, and if they follow suit in the technological transformation, they can combine their cheaper skilled workers and engineers with the new techniques to find a new basis for competitive strength.

Apart from the OECD countries, there is another group of countries that are benefiting from this technological transformation; namely those countries that do not have an engineering sector. These countries presumably get access to cheaper imported capital goods than would have been the case if the transformation had not occurred. The prime losers are probably those developing countries that have made a recent entry into the capital goods sector and have not yet adopted the new techniques. For these countries, the competitive environment in those products affected by flexible automation techniques has become tougher. Furthermore, the advantages that these countries may have in terms of cheaper labour are partly eroded.

As far as the more specific implications are concerned, it was mentioned above that the impact of flexible automation varies between different product groups. A number of products judged to be most affected by the new techniques were listed. The product groups were: automobiles, heavy electrical machinery, aircraft, cutting tools, pumps, valves, compressors, construction machinery, tractors, machine tools, special industrial machinery and electronic components. The impact also varies depending on how important relative factor prices/production costs are as determinants of international competitiveness in relation to other determinants such as firm-specific assets.

From an analysis of the structure of production and the trade in engineering goods in the NICs, a number of more specific conclusions were drawn:

1 The product groups that show the greatest impact of flexible automation techniques in the OECD countries constitute a large share of the value of production in the engineering sectors of the NICs.
2 These product groups are predominantly produced under import substituting schemes in the NICs. The Korean case shows this in a very clear way. The export performance of the NICs in these product groups is relatively poor and this is probably due to the fact that these products are technically complex and thus the determinants of international competitiveness go far beyond production costs. It cannot, however, be excluded that flexible automation techniques have already had an impact on the international competitiveness of the OECD countries at the expense of the developing countries.

3 Given the slow diffusion of flexible automation techniques in the NICs, their ambitions to become internationally competitive in those product groups which are presently produced under import substituting schemes has come up against a new obstacle, this being the improved competitiveness of the OECD countries on account of their rapid adoption of flexible automation technique.

4 For those product groups where the developing countries have shown a good export performance, predominantly technically simple product groups, the impact of flexible automation is relatively minor. The uneven global diffusion of flexible automation techniques therefore does not constitute a significant threat to the present export performance of the NICs in the engineering sector.

So what are the policy implications of all this for the developing countries? We would argue that, on the whole, the present diffusion of flexible automation techniques is much below the potential in the NICs. Hence, if the NICs could reduce the present obstacles to a faster diffusion of flexible automation techniques, at least some of the negative effects of this technological transformation would be mitigated. This applies in particular to NCMTs and CAD, which are of special interest to the developing countries because of their skill-saving biases. For these techniques, it seems that the main obstacles to a further diffusion do not lie in a lack of potential economic benefits from introducing them; instead, the obstacles lie in a lack of information about the techniques, restricted access to the techniques and a lack of knowledge of how to use them. The policy implications would then be that these obstacles to diffusion should be removed. More specifically, this means government actions in a number of areas, for example:

• To ensure that information is disseminated about flexible automation techniques.
• To change the educational systems so as to supply the right skills to use and maintain flexible automation techniques.
• To ensure that the would-be adopters have access to all the variants of these techniques and to after-sales service.

References

Alam, G., 1986: Interview by G. Alam with Marketing manager of ASEA's Transformer Division in January 1986. The result of the interview was transmitted orally to us.

American Machinist, 1983a: '13th American Machinist Inventory of Metalworking Equipment 1983', November.

—— 1983b: An FMS for Car-Engine Parts', December, p. 80.

Annuario estadistico do Brasil, 1982: Secretaria de Planejamento da Presidencie da Republica, Rio de Janeiro, Brasil.

Arnold, E., 1984: 'Computer aided design in Europe', Sussex European Papers, no. 14, University of Sussex, UK.

—— and Senker, P., 1982: 'Designing the future. The Implications of CAD Interactive Graphics for Employment and Skills in the British Engineering Industry', mimeo, Science Policy Research Unit, University of Sussex, UK.

Aronsson, R., 1986: *CAD använvondning i Brittish byggnadsindustri*. Utlandsrapport Nr 8603, Sveriges Tekniska Attachéer, Stockholm.

A:son-Stråberg, H., 1983: *Amerikansk Tillvedrkningsteknik – Utvecklingstendenser*. Utlandsrapport, Sveriges Tekniska Attachéer, Washington Ul-8301.

Association of Indian Engineering Industries, 1983: *Handbook of Statistics* (New Delhi).

Ashburn, A., 1981: 'A critical decision for America', *American Machinist*, November.

Bessant, J., 1983: *Technology and Market Trends in the Production and Application of Information Technology. A Review of Developments during the Years 1982–1983*. UNIDO Microelectronics Monitor no. 8, Supplement.

—— and Hayward, B., 1986: 'FMS in Britain – good and bad news', *The FMS Magazine*, January.

Boon, G. K., 1984: *Flexible Automation. A Comparison of Dutch and Swedish Firms, particularly as to CNC Machine Penetrations* (Noordwijk: Technology Scientific Foundation).

—— 1985: *Computer Based Techniques: A Technology/Policy Assessment in a North–South Perspective*. Noordwijk: Technology Scientific Foundation), Discussion Paper TSF 85-1.

Boston Consulting Group, 1985: *Strategic Study of the Machine Tool Industry*, Summary Report (Boston, Mass.: BCG).

References 213

British Robot Association (BRA), 1981: *Robot Facts – December 1981* (London: BRA).
—— 1982a: *Robot Facts – December 1982.*
—— 1982b: Press release from the British Robot Association, 9 February 1982.
—— 1983: *Robot Facts – December 1983.*
—— 1984: *Robot Facts – December 1984.*
Brown, Lawrence, A., 1981: *Innovation Diffusion. A New Perspective* (London: Methuen).
Brundenius, Claes, 1983: 'Some notes on the development of the Cuban labour force 1970–80', *Cuban Studies*, Summer 1983, vol. 13, no. 2.
—— 1984: *Revolutionary Cuba: The Challenge of Economic Growth with Equity* (Boulder, Col.: Westview Press).
Bryce, A. L., 1983: 'Is there such a thing as low-cost FMS?', in Rathmill (ed.), 1983, pp. 569–84.
Business Korea, 1985a: February 1985.
—— 1985b: September 1985.
Business Week, 1985: 16 June 1985.
Cambert, L. I., 1984: 'Robots as key-equipment in a highly mechanized body plant in the Volvo Car Corporation, Gothenburg', in N. Mårtenson (ed.), *Proceedings 14th International Symposium on Industrial Robots* (London: IFS (Publications) Ltd; Amsterdam: North-Holland).
Carlsson, Jan, 1983: *Production and Use of Industrial Robots in Sweden in 1982* (Stockholm: Computers and Electronics Commission, Ministry of Industry), Ds I 1983, p. 11.
—— and Selg, Håkan, 1982: 'Swedish industries' experience with robots', *The Industrial Robot*, June 1982.
Caves, R. E., 1980: 'Industrial organization, corporate strategy and structure', *Journal of Economic Literature*, vol. XVIII (March).
CECIMO (Comité Européen de Cooperation des Industries de la Machine-Outile) (Various years), International Statistics on Machine Tools, mimeo, Brussels.
Central Machine Tools Institute (CMTI), 1986: Machine Tool Census, Bangalore.
Chander, C., 1987: Interview with Mr C. Chander of DMC Computers Limited, New Delhi.
Chetty, N. 1982: 'Industrial robots: status and scope', *Electronics, Information and Planning*, April 1982.
Cheung, F., 1985: 'About NC machine tools in Singapore', Department of Mechanical and Manufacturing Engineering, Singapore Polytechnic.
Chudnovsky, D., 1984: *The Diffusion of Electronics Technology in Developing Countries' Capital Goods. Sector: The Argentinian Case* (Buenos Aires: Centro de Economia Transnacional).
—— 1986: 'The diffusion and production of new technologies. The case of numerically controlled machine tools', mimeo, Centre on Transnational Economy, Buenos Aires.
Cohen, Charmian (ed.), 1983: *The World Market for Industrial Robots* (University of Manchester: Centre for Business Research, Manchester Business School), May 1983.
Computers and Electronics Commission, 1981: *Datateknik i verkstadsindustrin* (Computer technology in the engineering industry), (Computers and Electronics Commission, Ministry of Industry, Stockholm, Liber Förlag), Sou 1981: 10.
—— 1983: *Investering i industrirobotar* (Investment in industrial robots) (Computers

and Electronics Commission, Ministry of Industry, Stockholm) DS I 1983:9.

Cooper, L. M. and Clark, J., 1982: *Employment, Economics and Technology* (Brighton: Wheatsheaf Books).

CSO (Central Statistical Office) (various years) United Kingdom National Accounts.

DEK (Data- och Elektronikkommittén), 1983: *Förslag till åtgärder avseende: Flexibla tillverkningssystem (PM1) Ingenjörsutbildning för industriarbetare (PM2)* (Stockholm: Ministry of Industry), DSI 1983: 27.

Dempsey, P. A., 1983: 'New corporate perspectives in FMS, in Rathmill (ed.), 1983, pp. 3–17.

Denis, F., 1983: 'Robots continue to join the Belgian workforce', *The Industrial Robot*, September 1983.

Design Graphics World, 1985: 'The shakeout begins – CAD/CAM/CAE industry growth slows to 30 percent', *Design Graphics World*, August, pp. 28–32.

DGTD (Directorate General of Technical Development) 1986: Machine Tools Cleared for Imports during 1984 (New Delhi).

Dupont-Gatelmand, C., 1983: 'Key success factors for FMS design and implementation', in Rathmill (ed.), 1983, pp. 283–93.

Ebel, K.-H., 1986: 'The impact of industrial robots on the world of work', *International Labour Review*, vol. 125, no. 1, January–February 1986.

ECE (Economic Commission for Europe), 1985: *Production and Use of Industrial Robots*. United Nations, Economic Commission for Europe, ECE/ENG.AVT/15.

—— 1986: *Recent Trends in Flexible Manufacturing* (New York: United Nations).

Economic Development Board (EDB), 1986: Interview with a representative of EDB, Singapore, November.

Economic Planning Board (EPB), 1982, 1984, 1985: Survey of Mining and Manufacturing Industry (Seoul).

Edquist, Charles, 1980: *Approaches to the Study of Social Aspects of Techniques*. Summary of a Doctoral Thesis (Lund: Research Policy Institute, University of Lund).

—— 1982: 'Technical change in sugar cane harvesting – a comparison of Cuba and Jamaica 1958–1980', Working Paper no. 96, World Employment Programme (WEP) (Geneva: ILO), July 1982.

—— (Geneva: ILO), 1983: 'Mechanization of sugar cane harvesting in Cuba', *Cuban Studies* (Center for Latin America Studies, University of Pittsburgh, USA), vol. 13, no. 2, Summer 1983.

—— 1985a: 'Technology and work in sugar cane harvesting in capitalist Jamaica and socialist Cuba 1958–1983', in Gustavsson, B., Karlsson, J. and Räftegård, C. (eds), *Work in the 1980s – Emancipation and Derogation* (London: Gower Publishing Company).

—— 1985b: *Capitalism, Socialism and Technology – a Comparative Study of Cuba and Jamaica* (London: Zed Books).

—— 1986: 'The impact on developing countries of flexible automation in the capital goods industry – a project outline', UNIDO, Sectoral Studies Branch, Division for Industrial Studies.

—— and Edqvist, Olle, 1979: *Social Carriers of Techniques for Development*, published as SAREC Report R3: 1979 by the Swedish Agency for Research Cooperation with Developing Countries, c/o SIDA, S-105 25 Stockholm, Sweden. (A somewhat abridged version under the same title was published in *Journal of Peace*

References 215

Research, vol. XVI, no. 4, 1979. In Swedish the study has been published as *Zenit Häften* 5, 1980.

—— and Jacobsson, Staffan, 1982: 'Technical change and patterns of specialisation in the capital goods industries of India and the Republic of Korea – a project description' (Lund: Research Policy Institute, University of Lund).

—— and —— 1984: 'Trends in the diffusion of electronics technology in the capital goods sector', Discussion Paper no. 161, Research Policy Institute, University of Lund, Sweden, August. (This study is partly the same as UNCTAD TT/65.)

—— and —— 1985a: 'Automation in the engineering industries of India and the Republic of Korea against the background of experience in some OECD countries', *Economic and Political Weekly*, 13 April 1985. Shorter versions have been published by WAITRO (World Association of Industrial and Technological Research Organizations): 'New technologies for developing countries: biotechnology and informatics', Stockholm, 1985, as well as in Advanced Technology Alert System (ATAS) II: 'Microelectronics based automation technologies and development', published by United Nations Centre for Science and Technology for Development, New York, 1985.

—— and —— 1985b: 'State policies, firm performance and firm strategies – Production of hydraulic excavators and machining centres in India and the Republic of Korea', *Economic and Political Weekly*, Annual Number, November 1985.

—— and—— 1986: 'The diffusion of industrial robots in the OECD countries and the impact thereof', paper presented at seminar on Industrial Robots '86 – International Experience, Development and Applications, Brno (Czechoslovakia), 24–28 February 1986. The paper has also been published in *Robotics*, vol. 3, no. 1, March 1987.

Ehnqvist, O., 1985: Interview with Olle Ehnqvist, Director, Marketing Division, Calma (International General Electric AB), Stockholm.

Elsässer, B. and Lindvall, J., 1984: 'Numeriskt styrda verktygsmaskiner. Effekter på produktivitet, arbetsvolym, yrkesstruktur och kvalifikationskrav', Ekonomiska Institutionen, Universitetet i Linköping.

Gebhardt, A. and Hatzold, O., 1974: 'Numerically controlled machine tools', in Nabseth L. and Ray G. E. (eds), *The Diffusion of New Industrial Processes: An International Study* (Cambridge: Cambridge University Press).

Giertz, Eric, 1983: *Den framtida verkstadsproduktionen* (Stockholm: Svenska Industritjänstemannaförbundet, SIF).

Goebel, R., 1983: 'Concept and realization of flexible manufacturing systems for machining of modular components', in Rothmill (ed.), 1983, pp. 69–79.

Gold, B., 1981: 'Technological diffusion in industry: research needs and shortcomings', *Journal of Industrial Economics*, vol. XXIX, no. 3, March 1981.

Gray, H. P., 1980: 'The theory of international trade among industrial nations', *Wirtschaftliches Archiv* (Tübingen) pp. 447–70.

Grumman Int. NTI CAD/CAM Centre, 1986: CAD/CAM Vision, March, (Singapore).

Halbert, B., 1985a: Statistics 1985-09-02. Föreningen Svenska Verktygsmaskintillverkare.

—— 1985b: Information received from Bo Halbert of the Swedish Machine Tool Manufacturers' Association.

—— 1985c: Telephone interview with Bo Halbert, Swedish Industrial Robot Association (SWIRA), August 1985.

Hansson, P., Karlsson, L. and Pärletun, L. G., 1985: *CAD/CAM/CAE – Datorstött ingenjörsarbete* (Lund: Studentlitteratur).

Hatvany, J., Merchant, M. E., Rathmill, K. and Yoshihawa, H., 1981: *Results of a World Survey of Computer-aided Manufacturing* (Washington, DC: National Academy of Sciences).

Helliwell, J. R., 1983: 'Flexible turning cells by SMT Machine Co AB Sweden', paper presented at Automan 83, 2nd European Automated Manufacturing Conference, Birmingham, England, 19 May 1983.

Hjelm, S., 1985: Telephone interview with S. Hjelm of the Institute for Production Engineering in Stockholm.

Hufbauer, G. C. 1966: *Synthetic Materials and the Theory of International Trade* (London: Duckworth).

Hull, S. M., 1983: 'A supplier's viewpoint of FMS', in Rathmill (ed.), 1983, pp. 537–46.

Hunt, T. L., 1984: 'Robotics, technology and employment', in *Proceedings of the 1st International Conference on Human Factors in Manufacturing*, 3–5 April 1984 (London: IFS (Publications) Ltd; Amsterdam: North-Holland).

Hägerstrand, Torsten, 1967: *Innovation Diffusion as a Spatial Process* (Chicago: University of Chicago Press).

IFS Publications, 1983: *FMS Update* (sample issue).

ILO (International Labour Office), 1981: *Year Book of Labour Statistics 1981* (Geneva: ILO).

—— 1982: *Year Book of Labour Statistics 1982.*

Industrial Robot International, 1982: vol. 3, no. 4, 8 March 1982.

Industrial Robot, The, 1981a: 'Robotics in the UK', *Industrial Robot*, March 1981.

—— 1981b: 'The state of robotics in Italy at the start of the 80's', *Industrial Robot*, September 1981.

—— 1983: 'Robots continue to join the Belgian workforce', *Industrial Robot*, September 1983, pp. 197–8.

—— 1984: 'Europe overtakes the USA', *Industrial Robot*, March 1984, pp. 38–41.

—— 1985a: 'Steady growth for robots in 1984', *Industrial Robot*, March 1985, pp. 30–5.

—— 1985b: 'Korea takes its first steps into robotics', *Industrial Robot*, June 1985, pp. 97–100.

Ingersoll Engineers, 1982: *The FMS Report* (Kempston: IFS Publications).

Jacobsson, S., 1981: 'Technical change and technology policy. The case of numerically controlled lathes in Argentina', mimeo, Research Policy Institute, University of Lund.

—— 1982: 'Electronics and the technology gap – the case of numerically controlled machine tools', in Kaplinsky R. 'Comparative advantage in an automating world', *IDS Bulletin*, vol. 13, no. 2. March 1982.

—— 1985: 'International trends in the machine tool industry', paper submitted to UNIDO and published as UNIDO/15.565, October 1985.

—— 1986: *Electronics and Industrial Policy. The Case of Computer Controlled Lathes* (London: Allen and Unwin).

Japan Economic Journal, 1986: *Japan Economic Almanac* (Tokyo).

Japan Statistical Yearbook (various years): published by Statistics Bureau, Prime Minister's Office.

Japan Tariff Association, 1985: *Japan Exports and Imports, Commodities by Country* (Tokyo).

Jones, D., 1983: 'Technical change and the Japanese challenge', mimeo, Science Policy Research Unit, University of Sussex.

KAIST (Korea Advanced Institute of Science and Technology), 1984: Interview with Lee Chong-Won and Ing Kom Moon Hyun at the CAD/CAM Research Laboratory.

Kamata, Shinichi, 1983: 'Robotization progresses swiftly amid "factory automation" boom', *Industrial Review of Japan/1983*, pp. 46–7, 69.

Kaplinsky, R., 1983: 'Computer aided design – electronics and the technological gap between DCs and LDCs', in Jacobsson, S. and Sigurdson, J. (eds), *Technological Trends and Challenges in Electronics* (Lund: Research Policy Institute), 1983.

Kim Linsu, 1986: 'New technologies and their economic effects: a feasibility study in Korea', mimeo, College of Business Administration, Korea University, Seoul.

Kjellberg, Torsten, 1984: Interview at the Swedish Institute of Production Engineering Research, Stockholm, January 1984.

Kobayashi, S., 1983: 'Economic efficiency still to be demonstrated', in *Metalworking, Engineering and Marketing* (Nagoya, News Digest), January 1983, p. 15.

Korea Machine Tool Manufacturers' Association (KMTMA), 1985, 1986: *Machine Tools* (Seoul, Korea).

Larson, L., 1985: 'FMS i Västtyskland'', Utlandsrapport från Sveriges Tekniska Attachéer, 8505.

Lindeberg, S. 1983: *Datorstött konstruktionsarbete. En kartläggning av CAD/CAM – tekniken och dess effekter*. Svenska Industritjänstemannaförbundet.

Livingstone, Gavin, 1981: 'The 1981 US robot industry', *Industrial Robot*, March 1981.

Lundström, G., 1987: 'Robotar ökar', *Ny Teknik*, no. 16, 16 April 1987.

McBean, D. J., 1983: 'Practical application of FMS', in Rathmill (ed.), 1983, pp. 477–84.

Machine Tool Trades Association (MTTA), 1983: *Basic Facts about the British Machine Tool Industry* (London: MTTA).

Machine Tool Trades Association and *Metalworking Production*, 1979, 1980, 1981, 1983, 1985: Machine Tool Statistics.

Mercado, A., 1984: 'La seleccion y difusion de maquinas – herra mienta de control numerico computerizado en Mexico'. The Technology Scientific Foundation, Documento de Trabojo TSF 84–6, Noordwijk.

Mertens, K., 1981: 'Entwicklungsstand Flexibler Fertigungssysteme in den USA', *ZWF*, 76 (1981), pp. 81–5.

Metalworking, Engineering and Marketing (Nagoya, News Digest), January 1982, May 1982, July 1982, September 1982, November 1982, January 1983, May 1983, July 1983, September 1983, January 1984, November 1986.

Metalworking Production, 1977: *The Fourth Survey of Machine Tools and Production Equipment in Britain* (London: Morgan-Grampian).

Moreau, L. Å., 1985: Interview with Lars Åke Moreau, Managing Director, Autodesk AB, Mölndal, Sweden, September 1985.

Moutrey, A., 1985: Letter from Alan V. Moutrey, British Robot Association, 10 April 1985.

Mueller, H., 1983: 'FMS: horizontal machining centers', in Rathmill (ed.), 1983, pp. 525–36.

Nabseth, L. and Ray, G. F., 1974: *The Diffusion of New Industrial Processes. An International Study* (Cambridge: Cambridge University Press).

Nagao, M., 1985: 'Automation technology diffusion within the engineering industry', *ATAS Bulletin* (New York issue 2), November 1985.

Nakao, E., 1983: 'Fanuc's factory automation (FMS)', in Rathmill (ed.), 1983, pp. 743–8.

National Machine Tool Builders' Association (NMTBA), 1981/2, 1982/3, 1983/4, 1984/5, 1985/6: *Economic Handbook of the Machine Tool Industry*.

Norström, C., 1983: 'The complete turning cell for FMS', in Rathmill (ed.), 1983, pp. 170–82.

Northcott, J., Rogers, W. and Knetsch, W., 1985: 'Microelectronics in industry, an international comparison: Britain, Germany, France', Policy Studies Institute Publication no. 635, London.

OECD (Organization for Economic Cooperation and Development), 1983: *OECD Economic Outlook*, 33, July 1983.

—— 1983b. *Industrial robots – their role in manufacturing industry* (Paris: OECD).

—— 1986. *OECD Economic Outlook*, 39, May.

Ohlson, L., 1976: *Svensk verkstadsindustris internationella specialisering.* Industrins Utredningsinstitut, Stockholm.

Palmer, L., Edquist, C. and Jacobsson, S., 1984: 'Perspectives on technical change and employment', discussion paper no. 167, Research Policy Institute, University of Lund.

Pang, A. H., 1985: 'Robotization of Singapore', mimeo, International University of Japan, Niigata.

—— 1986: 'Job creation-displacement analysis. The experience of robotization in Singapore', Paper presented at the 3rd International Conference on Human Factors in Manufacturing, 4–6 November, Stratford-upon-Avon, UK.

Perspective Plan Committee, 1983: 'The Indian Machine Tool Industry: A Perspective Plan 1983–93'. New Delhi.

Peterson, L., 1984: 'Svensk Utrikeshandel 1871–80. En studie i den intraindustriella handels framväxt', Lund Economic Studies, no. 30.

Popplewell, F. and Schmmoll, P., 1983: 'FMS from conception to action', in Rathmill (ed.), 1983, pp. 643–55.

Porter, M. E., 1980: *Competitive Strategy. Techniques for Analyzing Industries and Competitors* (New York: The Free Press).

Purdon, P. B., 1983: 'The Citroen flexible manufacturing cell', in Rathmill (ed.), 1983, pp. 93–103.

Rathmill, K. (ed.), 1983: *Proceedings of the 2nd International Conference on Flexible Manufacturing Systems* (Kempston: IFS Publications; Amsterdam: North Holland).

Rattner, H., 1984: 'La difusion de Maguinas-Herramienta de control numerico en Brasil', The Technology Scientific Foundation, Documenta de Trabaja no. TSF84–4 Noordwijk.

Rempp, H., 1982: 'Introduction of CNC machine tools and flexible manufacturing systems: economic and social impact', paper delivered to the Seminar on European Employment and Technological Change, Rome, 10–12 February 1982.

——, Botto, H. and Lay, G., 1981: *Wirtschaftliche und soziale Auswirkungen des CNC-werkzeugmachineneinsatzes*. Rationalisierungs-kuratorium der Deutscher Wirtschaft. RKW – Bestell no. 758.

Rizwi, N., 1985: Telephone interview with Mr Rizwi of ASEA.

Robot Application News, 1985: 'Robotiserad montering av vindrutetorkare hos Electrolux', *Robot Application News*, no. 3, October, p. 3.

Rogers, Everett M., 1982: *Diffusion of Innovations* (New York: The Free Press).

Rooks, R. W., 1986: 'UK robot application rate slows', in *Industrial Robot*, March 1986, pp. 3–4.

Rosegger, G., 1977: 'Diffusion of technology in industry', in Gold, B. (ed.), *Research Technological Change and Economic Analysis* (Lexington, Mass.: Lexington Books).

Rosenberg, Nathan, 1986: *Perspectives on Technology* (Cambridge: Cambridge University Press).

Sasaki, M., 1987: Lecture given by Professor M. Sasaki at the University of Lund, 20 May 1987.

SCB (Swedish Central Bureau of Statistics), 1983: Industri 1981, Stockholm.

Schumacher, D., 1983: 'Intra-industry trade between the Federal Republic of Germany and developing countries: extent and some characteristics', in Tharakan (ed.), 1983.

Sciberras, E. and Payne, M., 1985: *The UK Machine Tool Industry. Recommendations for Industrial Policy* (London: Technical Change Centre).

Senker, P., Swords-Isherwood, N., Brady, T. and Huggett, C., 1980: 'Maintenance skills in the engineering industry. The influence of technological change', mimeo, Science Policy Research Unit, University of Sussex.

SIRI (Sociata Italiana Robotica Industriale) 1983: 'Servo controlled robots installed in Italy', *SIRI Report*, January 1983.

Smith, Don, 1983: 'Delphi forecast of the future of robotics in the United States', speech presented at the 13th International Symposium on Industrial Robots, Chicago, 17–21 April 1983.

Soete, L., 1981: 'A general test of technological gap trade theory', mimeo, Science Policy Research Unit, University of Sussex.

Spur, G. and Mertins, K., 1981: 'Flexible Fertigungssysteme. Produktionsanlagen der flexibler Automatisierung', *SWF*, 76 (1981), no. 9, pp. 441–8.

Steinhilper, R., 1985: Communication by letter with S. Steinhilper of the Fraunhover Institute.

Steen, H., 1976: NC-inventerigen 1976, avslutande redogörelse by Harald Steen, Sveriges Mekanförbund, Avd. för Verkstadsteknik, 1976–10–21.

Swedish Industrial Robot Association, 1984: 'Robottekniken gör svensk industri ännu slagkraftigare', brochure published by the Swedish Industrial Robot Association, August 1984.

Tan, J. H., 1986: interviews with Mr Tan Joon Hong, Senior Industry Officer, Economic Development Board, Singapore.

Tauile, J. R., 1984: 'Microelectronics, automation and economic development', DPhil. thesis of the Ne School for Social Research, São Paulo, Brazil.

— 1986: 'The diffusion of microelectronics automated equipment in Brazil: economic and social implications'. The United Nations University, Rijksuniversiteit Limburg, Maastricht, November.

Tharakan, P. K. M. (ed.), 1983: *Intra-industry Trade* (Amsterdam: North Holland).

UNCTAD, 1982: *Problems and Issues Concerning the Transfer, Application and Development of Technology in the Capital Goods and Industrial Machinery Sector. The Impact of Electronics Technology on the Capital Goods and Industrial Machinery Sector: Implications for Developing Countries.* Study by the UNCTAD Secretariat, TD/B/C. 6/AC. 7/3, May, 1982.

UNCTAD TT/65, 1985: *The Diffusion of Electronics Technology in the Capital Goods Sector in the Industrialized Countries*, 11 September (G.E.85–57218). This study was prepared by Charles Edquist and Staffan Jacobsson in cooperation with the UNCTAD secretariat. Compare Edquist and Jacobsson, 1984.

UNCTAD TT/66 (forthcoming): *The Diffusion of Electronics Technology in the Capital Goods Sector: the Argentina case.*

UNCTAD TT/67 (forthcoming): *Issues Involved in the Introduction of Electronics Technology in the Developing Countries' Capital Goods Sector: the Yugoslav Case.*

UNIDO (United Nations Industrial Development Organization), 1985a: 'Conditions of entry into the capital goods sector and strategies for integrated manufacture', prepared by the UNIDO Secretariat for the Second Consultation on the Capital Goods Industry with Special Emphasis on Energy-related Technology and Equipment, Stockholm, 10–14 June 1985. ID/WG.442/1, 23 April.

— 1985b: 'Conditions of entry into the capital goods sector and integrated manufacture', prepared by the UNIDO Secretariat for the Consultation on the Capital Goods Industry with Special Emphasis on Energy-related Technology and Equipment, Stockholm, 10–14 June 1985. ID/WG.442/3, 26 April.

— 1985c: 'Report of the Second Consultation on the Capital Goods Industry with Special Emphasis on Energy-related Technology and Equipment', Stockholm, 10–14 June 1985. ID/338 ID1WG.442/5, 3 July.

United Nations, 1980: *Yearbook of Industrial Statistics* (New York: UN).

— 1982: *Statistical Yearbook for Asia and the Pacific* (New York: UN).

Utlandsrapport, 1984: 'Industrirobotar i Västtyskland', Utlandsrapport från Sveriges Tekniska Attachéer, September.

Utterback, J. M., 1979: 'The dynamics of product and process innovations in industry', in Hill, C. T. and Utterback J. M., *Technological Innovation for a Dynamic Economy* (Oxford: Pergamon Press).

VDMA, 1985: Information received from Verband Deutscher Maschinen- und Anlagenbau e.V.

Volkholz, V., 1982: 'Trends in the use of industrial robots in the 1980s – the case of the Federal Republic of Germany', *Microelectronics, Robotics and Jobs*, ICCP (Information, Computer, Communication, Policy) Series no. 7 (Paris: OECD).

Warnecke, H. J., 1983: 'New international developments for flexible automation in FMS', in Rathmill (ed.), 1983, pp. 681–95.

Watanabe, S., 1983: *Market Structure, Industrial Organisation and Technological Development: The Case of the Japanese Electronics-based NC-machine Tool Industry*. Technology and Employment Programme (Geneva: ILO), WEP 2–22/–WP 111.

Wiehe, R., 1983: 'Flexible machinery system', in Rathmill (ed.), 1983, pp. 337–46.

World Bank, 1982: 'Brazil: industrial policies and manufactured exports', report no. 3766–BR.

Yang, L. J., 1984a: 'The introduction of robot technology in Third World countries', mimeo, School of Mechanical and Production Engineering, Nanyang Technological Institute, Singapore.

—— 1984b: 'Applications of NC machine tools and flexible manufacturing systems for precision industries in Singapore', in proceedings of Seminar on Machine Tool Technology and Applications, Bandung, 1984.

Yonemoto, Kanji, 1981: 'The socio-economic impact of industrial robots in Japan', *Industrial Robot*, December 1981.

—— 1982: 'Robotization in Japanese industries – socioeconomic impacts by industrial robots', mimeo, Japan Industrial Robot Association, September 1982.

—— 1983: 'Robotization in Japanese industries – socioeconomic impacts by industrial robots', mimeo, Japan Industrial Robot Association, October 1983.

—— 1984: 'History and future outlook for robots in Japan', *Industrial Robot*, September, pp. 152–3.

—— 1985: Letter from Kanji Yonemoto, Executive Director, Japan Industrial Robot Association, 14 April 1985.

—— 1986: Letter from Kanji Yonemoto, Executive Director, Japan Industrial Robot Association, 21 August 1986.

Index

German Democratic Republic, robots
in 49
Ghandi, R. 164
Gold 13
Gold Star 154
government policy 18, 135, 154, 156,
186, 189, 196
effect on employment 121, 123–4
subsidy 158, 188
group technology 70–1, 96

Hägerstrand, T. 14
Hatvany, J. et al. 67
Hayward, B. 73
Hindustan Aeronautics 137, 159
Hitachi 68
Hufbauer, G. C. 108
Hughs Tool 137
Hungary, robots in 49
Hyundai Motor Company 154, 157,
158, 159, 166, 168

IBM 68
import restrictions 140, 149, 164,
165, 188, 206
import share of investment 179, 180,
181, 182, 206
import substituters 181–2
India 134, 149, 174, 178, 187, 209
CADs in 163, 164–5, 166
FMSs in 158–9, 208
NCMTs in 129, 130, 132, 138,
142, 145–6, 207
robots in 153
Indian Telephone Industries 164
Industrial Robot 155, 156
industrial robots 3, 4, 46–55, 74–6,
159, 198–9, 207–8
definitions 46–7
diffusion of 47–9, 153–5;
determinants 155–8
and employment 115–16, 118–19
industrial distribution of 50–3,
154
investment criteria for 53–4
as part of a FMS 62, 64, 65, 66
see also flexible automation

information, of new techniques 15,
16, 149, 155–6, 166, 186, 187,
196, 208, 211
diffusion of 12, 13, 14, 136–7, 206
institutional factors 9, 17, 34
interest, in new techniques 15, 16,
141–9, 156–8, 166–8, 196
International Harvester 68
investment shares in flexible
automation 5, 92, 94, 130, 175
islands of automation 3, 4, 97
Italy 47–8, 52–3

Jacobsson, S. 108, 120, 146
Japan 5, 23, 65, 73, 118, 133, 154, 156,
157, 175, 185, 203, 204, 208, 209
FMSs in 67, 68
NCMTs in 25, 26, 27, 28, 197
robots in 47, 49, 50, 51, 198
stock of flexible automation 114,
171–4
job creation 115–16; *see also*
employment; labour-saving
John Deere 69

Kaplinsky, R. 88, 167
KIA 159
Kia Motors 154
Kim Linsu 167
knowledge, of new techniques 15, 16,
138–41, 149, 155–6, 166, 186,
187, 188, 196, 199, 206, 208, 211
Kobayashi, S. 72
Korea 134, 175, 183, 184, 187, 189,
208, 209, 210
CADs in 163–4, 165, 166–8
FMSs in 159–60
NCMTs in 129, 130, 131, 138
robots in 153, 154–5, 156, 157–8,
185–6
specialization in 178–82
stock of flexible automation 171–3
Korean Advanced Institute of Science
and Technology (KAIST) 154,
166
Korean Institute of Machinery and
Metals 154